LANDS *of* LOST BORDERS

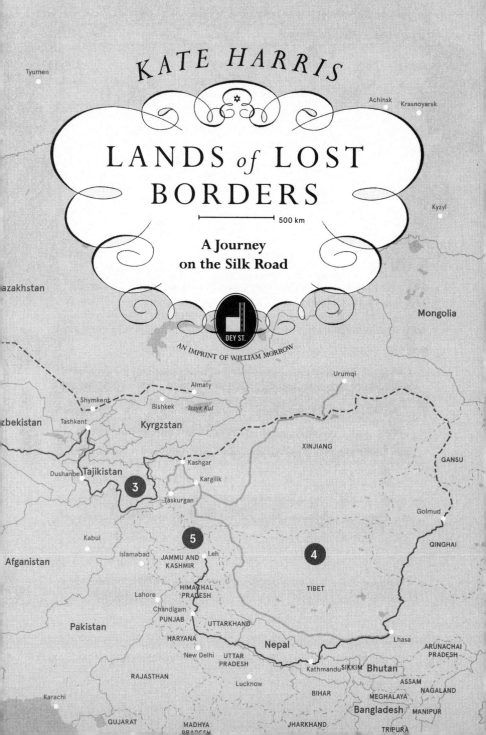

KATE HARRIS

LANDS of LOST BORDERS

500 km

A Journey on the Silk Road

DEY ST.

AN IMPRINT OF WILLIAM MORROW

Map by Five Seventeen

HarperCollins books may be purchased for educational, business, or sales promotional use. For information, please email the Special Markets Department at SPsales@harpercollins.com.

Published in 2018 by Alfred A. Knopf Canada.

FIRST U.S. EDITION

Library of Congress Cataloging-in-Publication Data has been applied for.

ISBN 978-0-06-283934-3

18 19 20 21 22   LSC   10 9 8 7 6 5 4 3 2 1

For my family, especially Nevs

To speak of knowledge is futile. All is experiment and adventure. We are forever mixing ourselves with unknown quantities.

VIRGINIA WOOLF, *THE WAVES*

CONTENTS

LANDS *of* LOST BORDERS

PROLOGUE

*T*he end of the road was always just out of sight. Cracked asphalt deepened to night beyond the reach of our headlamps, the thin beams swallowed by a blackness that receded before us no matter how fast we biked. Light was a kind of pavement thrown down in front of our wheels, and the road went on and on. If I ever reach the end, I remember thinking, I'll fly off the rim of the world. I pedalled harder.

The evening before, Melissa and I had carefully duct-taped over the orange reflectors on our wheels. Just after midnight, we'd crawled out of our sleeping bags, dressed in black thermal long underwear, packed up camp, and mounted our bicycles. As we rode toward Kudi, a tiny outpost in western China, only our headlamps gave us away, two pale flares moving against the grain of stars. We clicked off the lights as we neared the town.

It was three a.m. and moonless. The night air was cool for July and laced with the sweet breath of poplars and willows that grew in slender wands beside the river. No clean divisions between earth and sky, light and dark, just a lush and total blackness. I couldn't see the mountains but I could sense them around me, sharp curses of rock. The kind of country that consists entirely of edges.

Sometimes Mel and I drifted blindly into each other, our bulky panniers acting like bumpers. We navigated by the sound of our wheels, a hushed whirring indicating the pavement, a rasp of gravel the road shoulder and the need for a course correction. Travelling by bicycle is a life of simple things taken seriously: hunger, thirst, friendship, the weather, the stutter of the world beneath you. I was so focused on listening to the road that I didn't notice the glint of metal until Mel did.

"That's it," she whispered. "The checkpoint."

A guardrail scissored the road ahead, and somewhere beyond it, mythic and forbidden, was the Tibetan Plateau. Though Kudi isn't technically in the Tibet Autonomous Region, or TAR, as China has designated the formerly sovereign nation, the village hosts the first and most formidable military checkpoint on the only road into the western part of Tibet, a place foreigners require permits and guides to visit. Mel and I had neither. We didn't want to subsidize the Chinese occupation of Tibet by paying to go there, and we lacked the money for permits anyway. Plus, we'd just graduated from university and felt young and free and rashly unassailable: never once had we met a barrier we couldn't muscle past. So we took a deep breath, looked both ways, and biked directly under the raised guardrail.

Nothing happened. Somewhere to my left a river sounded like

wind. The stars looked freshly soldered above the dark metal of the mountains, faintly visible now that our eyes had adjusted. Mel was a whim of shadow to my left but I could feel her giddiness, or maybe it was my own, adding a kind of shimmer to the air. The world seemed preternaturally honed and heightened, our vision and hearing sharper. I watched a star shoot to the horizon with an afterimage trailing behind it. "Did you see that?" I whispered. When that same star shot up again, we shoved our bikes into the ditch and ran.

The flashlight scanned the road, moving closer in clean yellow sweeps. Mel dove into the ditch a few metres from our bikes and I bolted senselessly toward the nearest building, where I flattened myself against a wall. I heard footsteps approach, the click of heels on concrete, and regret seared me. I would never be a Martian explorer now. Instead I'd spend the rest of my days in a Chinese prison, desperately wishing I had something to read. With my cheek pressed against concrete, I stared up. If the heavens aligned, I told myself, if a single constellation clicked into place—the Big Dipper, say, or Cassiopeia—we'd be saved. I scanned the night sky for some reassuring sign, any familiar map to orient myself by— ironic, I suppose, when the great goal of my life was getting lost. But the stars reeled and spun and refused all their usual patterns. The footsteps came closer and closer and stopped.

Then I spotted the Big Dipper pouring out the sky. The foot- steps started again, moved closer, and faded away. I didn't dare move or breathe or glance at Mel, who was still playing dead some- where in the ditch. A few minutes or an eternity later a truck sput- tered into gear and drove off the way we'd come. The night settled back into silence.

We grabbed our bikes and continued racing through Kudi, instantly unrepentant. Fear exhausted itself into euphoria, a sense of irrational hope. The man with the flashlight surely saw us, pathetic and full of prayers in the ditch and against the wall, a couple of dogs with our heads tucked under the couch, believing our whole bodies hidden. At the very least he must have spotted our bikes overturned in the ditch, their wheels spinning uselessly. Why he decided to move on was a mystery we didn't question, in part because we were too winded to talk.

But even as Mel and I pedalled hard toward the Tibetan Plateau, I noted the bomb-like ticking of excess reflector duct tape against the front fork of my bike. *Tick-tick-tick-tick-tick*, the sound went, a gentle yet ominous stutter. *I should trim that*, I thought to myself. That's when a second checkpoint, the *real* checkpoint, loomed from the darkness like a bad dream. This time the guardrail was lowered, thigh-high, and secured with chains. Lighted concrete buildings edged the checkpoint on both sides, though we couldn't see anyone in them.

"Um . . ." I stopped pedalling, letting my bike coast and slow.

"Yeah . . . ," Mel acknowledged, but her voice came from somewhere ahead of me.

I hesitated for a beat and started pedalling again. If Mel wasn't about to back down, neither was I. "Throw your heart over the fence," our Pony Club instructors had always urged us, "and the rest of you will follow. Hopefully the horse and saddle too," they'd add with a grin. The only way to test the truth of a border is to ride hard toward it and leap—or, if circumstances demand it, crawl. Exposed in the pale light leaking from the checkpoint buildings, Mel and I glanced at each other one last time. Then we scuttled on hands and knees beneath the guardrail, dragged our loaded bikes after us, and pedalled as fast as we could into forbidden territory.

PART ONE

∿

How we spend our days is, of course, how we spend our lives.
ANNIE DILLARD, *THE WRITING LIFE*

1.

MARCO MADE ME DO IT

North America

*M*aybe all meaningful journeys begin with a mistake. Some kind of transgression or false turn or flawed idea that sets a certain irresistible odyssey in motion. Growing up in small-town Ontario, where the tallest summit was a haystack and the widest horizon a field of corn, my blunder seemed obvious, though it wasn't exactly my fault: I was born centuries too late for the life I was meant to live.

Restlessness runs in my family, though with my parents it mostly found expression in real estate. For the first decade of my life we lived in Oakville, a suburb not far from Toronto. But after spending their own childhoods mucking horse stalls and tending vegetable gardens, my engineer father and artist mother wanted the same rustic upbringing for my younger brothers and me, so when I was ten we moved to a few acres of cedar forest and swamp north of Ballinafad. This no-stoplight hamlet is a quaint tourist trap today, with the general store dealing in embroidered saddle pads and overpriced potpourri, but when I was a kid it was the kind of place even the school bus sped through to get somewhere else. When I was fourteen we moved again, this time southeast of Ballinafad, to a horse farm with seventy acres of woods and pastures, two spring-fed ponds, a barn full of empty boxes and shafts of dusty light, a log cabin so tiny I could almost touch any two walls at once, and a crumbling structure that once served as a sheep shed—but no house.

Somehow three restless kids, two patient adults, a barely house-trained Labrador puppy, and an indoors-only Abyssinian cat with an escapist streak crammed into a twelve-foot trailer for our first six months there, which was long enough to renovate the rundown sheep shed into a human dwelling of sorts. I say "of sorts" because it had a composting toilet instead of a septic system, a mouse once hitched a ride to school in my brother's backpack, and a snake slithered over my feet one spring when I was doing homework—details that mildly embarrassed my parents but delighted me, for they only enhanced the adventure of living there. "Race you to the sheep shed," I'd challenge my brothers as we spilled from the station wagon after a grocery run to town. "The *cottage*," my mom would correct me, insistent that the truth of a thing lay in its spirit, not the letter of its original design, but I was already off and running.

Compared to the trailer, the renovated sheep shed felt palatial at nine hundred square feet. I didn't even mind sharing a bedroom for a few years with my brothers. In our previous home, where I'd had my own room, I would always hear Dave and James chatting

and laughing through the wall, cracking jokes or doing imperson-
ations of teachers we shared over the years, like Mrs. Dingwall,
whose madcap name contrasted beautifully with her elegant
British accent, or Miss Pillon, a physics teacher who threw chalk
around the classroom to demonstrate the weak force of gravity,
thereby establishing for her students a lifelong association between
theoretical science and the instinct to duck. In the morning our
parents would find me cocooned in a duvet on my brothers' floor,
unwilling to miss the fun for the sake of a soft mattress.

With few kids our age nearby, the three of us had to entertain
each other. So we'd putter down to the pond on the lawn tractor,
hauling sand in the trailer to build a beach, until Dave backed a
little too close to the edge and the heavy load dragged the lawn
tractor into the water. Or we'd pull backflips for hours on our
trampoline, pretending we were on smaller planets, Pluto or Mars,
whose gravities didn't weigh us down as much. Then one winter
James tried to clear the trampoline of ice for some off-season practice and accidentally hacked through it with a pickaxe. We still
jumped on it for years, expertly avoiding the hole, until a visiting
friend ripped through it and put an end to our experiments in
soaring. After our grandmother informed us that we were related
to William Clark, of Lewis and Clark, we set off on rusty bikes to
pioneer a new route to the Pacific, stopping to resupply our expedition rations of red licorice at the Ballinafad general store.

But whatever direction we roamed, my brothers and I would
inevitably hit a wall. Sometimes it would be a fence, which we could
scramble over, but more often a highway or cookie-cutter housing
complex, paved and implacable, would stop us dead. The older I
got, the more our neighbourhood began to feel quaint and delimited, more rustic than rugged. Dave and James, three and five years
younger than me, didn't seem too bothered by this. They were just

as happy indoors, where they would construct model *Star Trek* spaceships or compose songs on the synthesizer my dad built. But the tamer my surroundings, the more I began to crave the antithesis: deserts and polar tundra, mountains and glaciers. The windswept margins and the steepest verges. The kind of wildness that could wipe me out if I wasn't equal parts bold and careful. In southwestern Ontario, I mostly found it in books.

My literary tastes, like my imaginative life, tended to the alien and extreme. Between homework and mucking horse stalls, on the school bus and at the dinner table—until my parents threatened to withhold dessert if I didn't put the book down—I wandered the Empty Quarter with the Bedu, searched Cape Royds for a penguin's egg, slogged east to west across Greenland on wooden skis, snapped photos of the dark side of the moon, answered the call of the wild in the Yukon, and trespassed across the Tibetan Plateau disguised as a Buddhist pilgrim. "I have a homesickness for a country that isn't mine," Alexandra David-Néel wrote about her stealth journey across Tibet, a country even more restricted to foreigners in 1924 than it is now. "The steppes, the solitude, the eternal snows and big skies up there haunt me."

David-Néel's book about that expedition, *My Journey to Lhasa*, was the closest I'd found to a portrait of the explorer as a young woman. Never mind that she was fifty-five when she donned her sheepskin cloak and trespassed boldly into Tibet (accompanied by her adopted Tibetan son, Yongden); age was less relevant to me than motivation. David-Néel wasn't trying to "find herself" through travel. Nor was she jolted from a routine, domestic existence by some kind of emotional crisis, as though only grief or loss or a search for love could justify a woman seeking risk and adventure on the open road. Refreshingly, David-Néel knew herself just fine, and what she was searching for, if anything, was an outer world as wild as she felt within. She didn't even have the luxury of a blank literary or geographic slate when it came to Tibet: dozens of Europeans had already been there, from diplomats to missionaries to soldiers. They'd drawn maps, written reports, even owned real estate in Lhasa. That none of this deterred the Frenchwoman was deeply consoling to me, a hint that exploration was possible despite precedent, that even artificial borders were by definition frontiers, and therefore worth breaching as a matter of principle. What propelled David-Néel onto the plateau was her wide-cast sense of wonder, exuberant wiliness, and fondness for travelling under the stars by night—in part to avoid being caught by day. In her era, Tibetan officials, not Chinese police, were the authorities to evade.

Tibet first cast a spell on me at an even younger age, maybe ten or eleven, when I found an illustrated, abridged edition of Marco Polo's travels on the Silk Road, the ancient caravan route that for thousands of years ferried people, goods, creeds, and ideas between Europe and Asia. The book had been my mother's as a child, and I loved seeing her maiden name elegantly inscribed on the inside cover, as if endorsing the adventures contained within. Its pages showed the seventeen-year-old Polo roaming far-flung lands with a camel caravan in tow, gazing at horizons that melted into fantastic mirages—turquoise-tiled domes and shifting deserts, labyrinthine bazaars and ice-mazed mountains. Polo looked bold and rugged and every bit the intrepid explorer. I decided to be just like him when I grew up.

Meanwhile I plotted his travels across the pages of an atlas, tracing the Silk Road, which actually consists of many roads, as it laced and frayed past Constantinople, Trabzon, Erzurum, Bukhara, Samarkand, Badakhshan, Kashgar, Khotan, Cathay, each name an invitation to elsewhere. But even more compelling, then and still, were the hinterlands between those trading hubs. Not only the

Tibetan Plateau, that upheaval of rock and ice and sky, but also the Pamir Mountains, where herds of sheep with improbably huge horns dodged avalanches and snow leopards with an elegance close to flight. And the Taklamakan, a shifting sands desert dwarfed only by the Gobi and Sahara whose name, according to legend if not literal translation, means "he who goes in never comes out."

I would've gladly gone where none had before, with no promise of return, for even a whiff of insight into the basic perplexities of existence: Where did we come from and are we alone in the cosmos and what exactly—or even generally—does it all mean? Places like the Tibetan Plateau or the Taklamakan Desert seemed to promise not answers, exactly, but a way of life equal to the wildness of existing at all. Even more compelling than far-flung mountains and deserts were the stars above and beyond them, distant suns lighting who knows what other worlds. Only I couldn't imagine how to reach them: the Voyager I and II spacecraft were long gone by the time I was born.

Launched by NASA in 1977 to study the most distant planets in the solar system and then cruise forever into interstellar space, the Voyager probes were the farthest human-made objects in the universe when I learned about them in my eighth-grade science class. I got chills thinking about those robotic emissaries speeding out past the heliopause—the outermost boundary of our solar system—into the largest possible story of what is. What would they see out there? Who would they meet? How could we stand to never know, given the difficulties of data transmission across galaxies?

I would've jumped at the chance to hitch a ride on either of the Voyagers, their lack of life support systems notwithstanding. Of course I would ache for family and friends, setting off for some faraway place with no escape route or ticket home. I'd miss my books and my brothers and even the sheep shed. But it was the truth I was after, the deepest wonder, nothing less. "The the," wrote Wallace Stevens in a poem I read years later. I was grateful someone had finally managed to articulate it.

The word *desire*, at its root, means "of the stars," which seemed self-evident by the time I reached high school. After studying all the atlases I could find, I'd concluded with a sense of panic that I was wilder than the world in all directions. My neighbourhood wasn't the only place circumscribed by an expanding network of highways and subdivisions; most of the planet was similarly under siege. My family couldn't afford to travel abroad, and I worried that by the time I'd saved enough money to see Tibet for myself, it would be as tame as Ballinafad. There seemed few outlets left for the restlessness that ached inside me, this mad longing for a world without maps. My only hope, I realized eventually, was to leave the Earth behind. So I wrote a letter urging a human mission to Mars and mailed it to twenty-two world leaders.

"I am a seventeen-year-old girl who has a dream," I declared to Bill Clinton, Tony Blair, Jean Chrétien, Jacques Chirac, and other influential heads of state in 1999. "That dream is for humanity to go to Mars."

Why the red planet, given the plurality of possible worlds? Because human physiology is as fussy as Goldilocks, and most planets are too hot, too cold, too big, or too gaseous to be habitable. Mars, if not just right because of its poisonous, fatally thin atmosphere, is otherwise fairly close: a world roughly the consistency and size of our own, only with a day that lasts longer by twenty-nine minutes due to its slower planetary spin, and a weaker gravitational pull due to its smaller mass. The gift of time, a lighter step—what wasn't there to love? With gorges five times deeper than the Grand Canyon, deserts many times drier than the Taklamakan, and a mountain triple the height of Everest, Mars is a world of geological superlatives—and exploratory firsts waiting to happen. And while the red planet might lack little green men, little green microbes are a genuine possibility, given that single-celled organisms can survive similarly cold, dry conditions on Earth. Mars is also pocked and scarred with features that hint at a warmer, wetter past, when conditions might've been more clement for life as we know it. That neighbouring world, in short, could well supply an answer to the age-old inquiry "Are we alone?"

In my manifesto, I rhapsodized to world leaders about how the urge to explore the unknown is ingrained in the human spirit. I reasoned that we had all the technology we needed to send humans to Mars, and all we lacked now was political will. I explained that the knowledge we could potentially gain there, such as proof of alien life, could have immeasurable benefits for people on Earth, such as making us feel less lonesome. I stressed that such a journey would ignite the passions of the world's youth. "It is the inquiring minds of bold dreamers and explorers such as Magellan and Copernicus that help to extend the boundaries of knowledge, enabling the human race to understand more, to see further," I wrote, noting that a human mission to Mars was the modern equivalent to these historic voyages—and an enterprise worthy of association with their good names.

In reply I received a few desultory form letters. But if my missive didn't launch a new era of interplanetary exploration, it did win me the Hakluyt Prize, given by the Mars Society for the best student letter advocating the human exploration and settlement of Mars. My reward was an eight-inch Bushnell telescope through which, late one night on the lawn outside the sheep shed, my father helped me spot the rings of Saturn for the first time. I also won an all-expenses-paid trip to the International Mars Society Convention.

Most teenagers long for another world, but as far as I could tell in Ballinafad, I alone pined specifically for Mars. The convention, held that year in Boulder, Colorado, upended this feeling of isolation. I stood on a podium and shyly read my manifesto to an auditorium full of scientists, engineers, and other anachronistic explorer types who'd perhaps found themselves stranded, like me, on a depressingly fenced-in, paved-over planet. Academy Award–winning filmmaker James Cameron was among them, and Apollo moonwalker Buzz Aldrin. They gave me a standing ovation, just a sweet gesture of encouragement to a kid, but in that moment of being heard, I felt unlimited. These were my people, I exulted. Here was my tribe. I vowed to become a scientist and go to Mars.

Science had long been my favourite subject at school, and not just because of the red planet. Science fair projects were a grand excuse for several weekend sleepovers in a row with my best friend, Melissa, who lived nearly an hour away. Other than at Pony Club, which only met in the summer, I rarely got to see her outside of school. In the sixth grade the two of us tested whether human saliva was more bacterially diverse (read "disgusting") than dog drool—an experiment that began as a ploy to embarrass our younger brothers, whom we duped into donating spit. We won a medal at the science fair, though not without a few raised eyebrows from the judges, and I marvel that I ever thought I had a future in microbiology.

Blame the microscope I got for Christmas when I was about thirteen. This gift from my parents was less a scientific instrument, I quickly realized, than a way of seeing everything as if for the first time. Ordinary, everyday things—the cuticle of my thumb, a drop of pond scum—looked alien upon closer scrutiny, with unmapped mountain ranges and nameless oceans swarming with life. My stomach would flip as I stared into the distances of a single-celled alga, whose long Latin name and twitchy, see-through form seemed proof that life was just as I'd suspected it all along: a mystery we can barely pronounce, never mind keep entirely in focus. In high school a few years later, Darwinian evolution put a whole new spin on existence for me, as though I'd been staring at a portrait of biology for years upon years, studying its eyes, ears, and nose, mapping the pores and wrinkles of its face, and in a flash suddenly grasped its *expression*. Learning about Darwin's seven-year voyage on the *Beagle*—in which he sailed around South America, collected strange life forms, and began formulating the theory of evolution by natural selection—taught me another valuable lesson about science: you can hitch a ride on it to some truly far-out places. So when the Morehead-Cain Foundation offered me a full scholarship to study biology at the University of North Carolina at Chapel Hill, I accepted it over the phone without hesitation, despite never having seen the campus and knowing nothing of the American South. The scholarship came with summer travel grants, and that was all I needed to know. I was desperate to see the Tibetan Plateau or the Taklamakan Desert in more than pixels or words on a page.

Even before the travel grants kicked in, the scholarship paid for sturdy hiking boots and a twenty-eight-day Outward Bound course in Utah the summer before my freshman year. Until then I'd only car-camped on family vacations to Ontario provincial parks, and despite my voracious reading and Ballinafad-based adventures, serious expedition travel seemed daunting to me, requiring technical skills and equipment I didn't have. Utah was a revelation: I learned how to slog up mountains and across deserts, carrying a fifty-pound backpack crammed with all I needed to survive— mainly oatmeal, a tarp and a sleeping bag, and a secret stash of books. I learned how to read where desert springs might be found in the contours of maps, and failing that, how to salvage drinking water from rain puddles afloat with dead frogs. I absorbed so much red dust into my pores that I began to resemble the red planet. On a daily basis the rough-hewn wonder of that place and experience brought me to my knees, in every possible sense. It was torture. It was sublime. It was basically everything I'd ever wanted.

I spent my next four summers slogging across similar immensities of stone and sky, toting along my weight in books. I stretched those scholarship travel grants as far as they would go, which, as it turned out, was pretty much anywhere short of Mars. Though funding wasn't guaranteed even to scholarship students, if I mustered a good enough reason for exploring a place, and justified my reasons for going in a written proposal, I could practically issue my own ticket. And so I stalked Sumatran rhinos in the jungles of Borneo, tracked wild horses through the Gobi desert in Mongolia, and skied across the Juneau Icefield for a glaciology field course, which I enrolled in never having seen a glacier and finished wanting to see little else. That cold spill of ice, rock, and sky bordering Alaska and British Columbia sent me skidding with its splendour, or maybe that was my lack of skill on skis. Either way, my preferred method of procrastination back at university became Googling for cabins for sale in Atlin, the remote BC town in which the glaciology course finished. I also became obsessed with Antarctica, a continent I envisioned as the Juneau Icefield on steroids—but travelling there would cost far more than a scholarship grant could cover.

This was when I realized I could still issue my own ticket if I pitched the right idea to the right people at the right time, in language so compelling and persuasive they couldn't possibly say no.

And so I turned to writing during university less out of a love for words (though there was certainly that) than for where they might launch me, such as the McMurdo Dry Valleys of Antarctica, where I went as a research assistant to a brilliant, generous scientist I politely lobbied for years. With a similar level of dedication, study, training, sacrifice, and more lobbying in the form of a written application, I also managed to launch myself to Mars—or at least to Hanksville, back in Utah, a place nearly as red and remote from human concerns. Once the desert hideout of Butch Cassidy and the Wild Bunch, who shook off law enforcers in a maze of red canyons, Hanksville now hosts crews of spacesuit-clad scientists and engineers on two-week simulated Mars missions. Picture ATV tracks, clumps of sagebrush, and a white space capsule gleaming cinematically before a world gone to rust.

For a while it was fun, a grown-up game of make-believe. Yet as four crewmates and I trundled around Utah in canvas spacesuits, I found myself disconcerted by the fact that when I gazed at a mountain, I saw a veneer of Plexiglas. When I reached out to touch canyon walls the colour of embers, I felt the synthetic fabric of my glove instead of the smooth, sun-warmed sandstone. As all kinds of weather howled outside my spacesuit, I heard either radio static or my percussive panting amplified in the plastic helmet, like I was breathing down my own neck. The very technologies that would sustain me on Mars made me feel at a deep remove from the place, my interactions with it neutered and sterile and more than slightly absurd.

"Okay, crew," ordered Commander Roger, a fifty-something engineer. "Fan out and find us some fresh rations!"

Having run out of food on the red planet, we patrolled the aisles of the local grocery store, wearing spacesuits to avoid fully "breaking sim," the worst misdemeanour aspiring Martian colonists could commit. I headed for the vegetable aisle with Tiffany, a molecular biologist, while another engineer named Allan and a geologist named Shahar hustled toward freezers of hamburger meat. When I glanced over at Gernot, an astronomer, he was frozen in front of the beef jerky stand, his helmet fogging over the flavours he couldn't find in his native Austria: honey-glazed, salt-and-pepper, teriyaki, chipotle, mesquite barbecue.

"Move along, Gernot," scolded Roger. "I said *fresh*."

After half an hour we regrouped at the checkout counter, our arms spilling with terrestrial bounty. The middle-aged cashier had seen our ilk before. "How's space camp treating you?" Roger visibly bristled inside his slightly too-tight canvas spacesuit. "It's not space camp," he shouted through his helmet. "It's a Mars simulation!"

"Easy, buddy," soothed Allan, patting Roger on the back. Gernot, seeing his chance, slipped two bags of beef jerky onto the checkout counter. Tiffany flipped through a celebrity tabloid and pretended not to know us, as if she were just another tourist dressed like Sally Ride. An older lady shuffled into the store, saw us, and hastily shuffled back out.

"So," the cashier said as she bagged the last of our rations, "I assume NASA will want a receipt for this?"

Maybe my own motives weren't so different after all from Cassidy's Wild Bunch: to escape a given reality, to flee to less mapped and lawful territories, to go rogue. Strangely, my crew-mates didn't seem to miss anything on Mars, not the lack of fresh air or birdsong, not the freedom to dictate our own days. Then again, the ideal Martian colonist doesn't complain. In fact, the ideal Martian colonist must possess a deeply paradoxical blend of personality traits: They should be emotionally astute and empathetic in order to thrive in a small social group under stressful conditions, yet detached enough from life on Earth to leave it behind forever. They must be sufficiently daring and rebellious to edge into realms no one has risked before, but not so independent or dismissive of authority that they don't obey orders once they get there. A gregarious recluse, in other words, a compliant adventurer. I used to think I was the perfect candidate.

But after two weeks of following orders, speaking in acronyms, and inhaling recycled air, I'd had my fill of living in a bubble, even the red-tinted variety. I never admitted it to the rest of the crew, but I was homesick for my native planet. So on the final night of the simulation, when everyone was asleep, I slipped out of the airlock without donning my plastic helmet or canvas spacesuit, without confirming my plans with Mission Control, without a radio to report back on my every sneeze. Crossing the equivalent threshold on Mars, I would've died in so many ways at once: poisoned, flash-frozen, depressurized. But here the Earth simply announced itself with a gust of wind spiked with sage, beneath a night sky tacked up with stars.

The first sign of doubt is a renewed fanaticism. Back at university I studied harder than ever in hopes of becoming an astronaut, despite questioning, deep down, whether I wanted to permanently emigrate to Mars if it meant a lifetime of containment. I spent my spare time presiding over a space club and volunteering in a marine microbiology lab, where my job was to tease long, invisible threads of DNA from the basaltic crust of the Pacific Ocean. Though I didn't have the opportunity to retrieve those sea-floor samples in a submarine, the lab was located in a windowless basement, which offered a similar experience of sensory deprivation only without any of the adventure. Whenever I walked outside, pale and blinking, I felt as if I'd been submerged for months. Books kept me going like bubbles of oxygen. One afternoon, shielding my face against the sun after the dim fluorescent flicker of the lab, I settled onto the grass on campus with my old friend Marco Polo, hoping to let my eyes re-adjust to wider horizons with the help of a childhood hero—this time in his full, unabridged glory.

But as I read *The Description of the World*, I was shocked to encounter a stranger in the Venetian explorer, someone who didn't relish slogging across lands that left me dizzy with longing. Instead, this Polo skirted the Taklamakan's wandering dunes as widely as possible, meekly plugging his ears against the spirit voices he feared would lure him into the trackless sands. He may never have seen a big-horned sheep alive in the Pamir, just their horns carved into bowls or stacked as fences, though this species, *Ovis ammon polii*, was eventually named after him: the Marco Polo sheep. And when he visited the Tibetan Plateau, he dismissed it as a blighted wasteland. "You ride for twenty days without finding any inhabited spot," he complained, "so that travellers are obliged to carry all their provisions with them, and are constantly falling in with those wild beasts which are so numerous and so dangerous." Again and again the so-called explorer dashed around mountains and deserts as quickly as possible, cursing wilderness as a mere obstacle to swift progress and profit.

I couldn't exactly blame him, for the Silk Road was a thirteenth-century superhighway for trade, after all, and Polo a merchant. After following his businessmen uncles to Cathay, where Kublai Khan took a shining to the Venetian teenager, Polo was tasked with assessing the value and variety of goods throughout the Mongolian empire, which at the time spanned Asia to the edge of Europe. Polo took this duty seriously, and as a result his travelogue reads more like a catalogue. The book details the precious commodities available along the Silk Road: silver in Armenia, rubies in

Badakhshan, black magic charms in Kashgar, ivory in India. With regard to lands less obviously exploitable, such as Tibet, the text is far terser.

I was gutted. Like so many explorers falsely portrayed as noble trailblazers in my high school history textbooks—from Christopher Columbus to Sir John Franklin—Polo turned out to care mostly about fortune and fame. Did no one but Alexandra David-Néel set off for the pure sake of setting off, propelled by a basic need to see around the bend, without the ulterior motives of wealth and conquest? Sitting in the quad, feeling bereaved, I resolved to see the Silk Road for myself, on a pilgrimage to the precise wildernesses Marco Polo most feared and shirked. I briefly considered travelling on horseback, but the lack of water in the Taklamakan was concerning. Camels were better suited to such arid environments, but the terrain would be bumpy enough as it was. A bicycle struck me as the perfect substitute: self-propelled and unlikely to spit.

That night I called my parents and outlined my preposterous plan. There was a long silence on the line. Then my mother said, "Please go with a friend?"

So I asked Mel if she wanted to join me on a bike ride. Just the Chinese section of the Silk Road for starters, I reassured her; it boasted the greatest concentration of places Polo most dreaded. After a warm-up trip across the continental United States that summer, we graduated from university the following year and hit the Silk Road. And that's how we found ourselves dodging landslides in the Pamir Mountains, gritting teeth through sandstorms in the Taklamakan Desert, and slinking toward Tibet while the stars looked the other way.

2.

ROOF OF THE WORLD

Tibetan Plateau

Salvaging the art of exploration, however, wasn't so simple. When Mel crawled beneath the checkpoint guardrail in China, she prudently left ample clearance between her back and the metal rod. In my haste and terror, I didn't scurry quite low enough. I'm not sure whether my backpack or my helmet snagged against the metal guardrail, rattling the chains that secured it, but either way I might as well have sounded a gong. Dogs barked, lights blazed, a voice shouted into the night—but we were already gone, racing into a tar-like darkness. We were mired in it, we couldn't pedal fast enough, we couldn't see anything but stars. I nearly lost control of my bike when I rode blindly into a pothole, and soon after a pylon. At the first hint of pursuing headlights I was ready to abandon my bike and flee up a mountain or into the river. But after minutes, and then hours, none appeared.

I felt the relief first in my fingers, unclenching from the handlebars, and next in my legs, which turned to slush. Technically the TAR was more than a hundred miles and several high passes away, but Kudi was the biggest bureaucratic hurdle on the way there. Because the checkpoint was in a narrow valley next to a roiling river, the Chinese authorities tended to assume people couldn't sneak around it, which meant Mel and I could breathe a little easier on this side: if anyone saw us they'd assume we had permission to be there. Unless they happened to be police from the checkpoint, chasing us down.

Eventually dawn lit the land around us, revealing mountains as rough as gnawed-down fingernails. The ragged peaks stretched on as far as I could see, a fury of forms. Rock turned to rust in the low-angled light and faded to umber and grey as the sun rose higher. A flock of dusty birds I couldn't name swooped above the river, whose turbid surge was distilled at this higher altitude to a clear stream, its water no longer the colour and texture of chocolate milk. I felt thin and insubstantial as a shadow, but the day had barely begun. Around every bend in the road I braced myself for a police convoy, a glimpse of the plateau, a woolly mammoth. Nothing would've surprised me, for the world seemed less unknown than unknowable, wavering around me like a half-formed thought. Then I realized I was dizzy with thirst.

I reached down for my water bottles, but the first was empty and I couldn't find the second—probably lost in the turmoil at the checkpoint. I told Mel to continue on while I stopped to fill my bottle in a roadside stream. Because of the steady drawl of the water

I didn't hear the car pull over. I turned around and there it was, puttering with menace, some sort of government emblem on the door. When a chubby Chinese man in a crisp navy blue uniform got out I knew it was over, for the third time that morning.

Without saying a word, the Chinese cop kicked my bike's tires and tried to lift its frame. The heavy bike hardly budged. Shaking his head, he returned to the car and fumbled in the trunk. For an arrest warrant, I was sure, possibly handcuffs. Instead he returned with three crisp cucumbers.

"Hello!" he grunted as he handed me the vegetables.

"Oh," I said, stunned. "Thank you!"

Without another word he got into the vehicle and drove off.

I caught up with Mel, who had been oblivious to my plight, and gave her a cucumber. She looked surprised, but a cyclist is never one to turn down a snack. We continued biking, munching as we pedalled, and by midday reached the bottom of a ten thousand–foot pass, the first step on the hypoxic staircase of passes climbing onto and across the Tibetan Plateau, where the average ground elevation is nearly as high as Mont Blanc. Lacking the energy and nerves to tackle the pass that day, we found a gully deep and wide enough to camp in and lounged there all afternoon, trying to ignore the imminent prospect of discovery by the Chinese police. Cucumber Cop had probably told his colleagues there was no need to rush, convinced we couldn't get far on such heavy bikes.

Instead, that afternoon, we were discovered by our new friends. I'd met Ben the previous summer at a hostel in San Francisco. After learning he was a bike mechanic, I casually invited him to join Mel and me on a cycling trip in China the following summer, and to our surprise, he did. We'd met Florian and Mattias, two German cyclists, in a hotel in Kashgar, and we'd biked as a peloton until a few days earlier, when Mel and I hit the road early and took an impromptu nap in the shade. We assumed the boys would catch up, see us dozing on the side of the road, and serve as our alarm clock. Instead, we woke at dusk not knowing if they were ahead or behind.

After the checkpoint, we never expected to see them again. In fact, they'd almost ridden past our hiding spot when Ben spotted a curl of Mel's red hair among the rocks. Confusing the ruddy flash for a camel, he stopped for a closer look and found us instead. Once we were reunited, Mel and I told them about our crossing—the truck driver! The shouts! The blind and desperate breakaway! Then we listened to their version.

"We scoped out the checkpoint from a distance during the day," said Ben. "Like you, we'd planned to go through at night, but then we saw these rabid-looking guard dogs!"

"Me, I don't like dogs," declared Mattias, a thick Bavarian accent making his every pronouncement sound profound.

"So in broad daylight we biked right up to the checkpoint—" said Ben.

"Showed our passports to the guards—" added Florian.

"And they waved us across," finished Ben with a smirk. "No questions asked."

The higher we climbed onto the Tibetan Plateau, the better I could breathe. I felt a strange lightness in my legs, an elation of sorts. Each revolution of the pedals took me closer to the stars than I'd ever propelled myself, not that I could see them by day, when the sky was blue and changeless but for a late-morning drift of clouds. The shadows they cast dappled the slopes of mountains like the bottom of a clear stream, so that climbing the pass felt like swimming up toward the surface of something, a threshold or change of state. Earth to sky, China to Tibet.

My tires scrabbled for traction on the loose knuckles of gravel paving National Highway 219, the only road leading into and across western Tibet. After just two switchbacks we were high above our last camp, and I could see Ben and the Germans milling around below, dawdling as usual. Mel and I preferred waking and biking early, when the land came alive in the slanted light of morning and it seemed we had time enough to get anywhere by nightfall, Lhasa or the moon. Florian, Mattias, and Ben preferred to sleep late, boil enormous pots of sweet milky rice for breakfast, and amble onto the road at midday. We usually crossed paths again in the late afternoon, when they either caught up with us or found our camp.

Mel and I biked up the pass side by side, barely speaking, sent into parallel solitudes by the effort of the climb. I'm not sure where I go when I spin wheels for hours on end like that, except into the rapture of doing nothing deeply—although "nothing," in this case, involves a tantrum of pedal strokes on a burdened bicycle along a euphemism for a highway through the Himalaya. But in the singular focus of that task, the almost tantric simplicity of it—breathe, pedal, breathe—I took in everything at once: the dust settling on my skin, the ache and strain and release of my quads, the river glittering far below like an artery of light, a shining silver vein, surely not the same sludge-like flow we'd camped next to a few days ago. Ride far enough and the world becomes strange and unknown to you. Ride a little farther and you become strange and unknown to yourself, not to mention your travelling companion.

"Nice face mask, bud," Mel managed between pedal strokes. "Wearing enough sunscreen?"

I grinned through a thick mask of sweat and grit and sunscreen, which I never rubbed in upon application, convinced it worked better as an opaque, unabsorbed gloss.

"You're one to talk!" I told Mel. "Your hair is growing its own hump."

I can't remember exactly how we became friends, but I believe it had something to do with volleyball. When my family moved north of Ballinafad, at the age of ten, I was the bookish new kid in a school where Mel was universally adored for her unconstellated freckles and the red hair she hated, for her sidelong sense of humour and winning habit of throwing her head back when she laughed. We had little in common until gym class, where the two of us were among the few kids who dove to stop every spike, no matter how futile the reach, how unforgiving the floor tiles. Our team went on to lose every match for the next three years of elementary school—not just every game, but every set. I didn't mind, and neither did Mel. The point of life, by our mutual measure, was to give it all we had. The only way we knew how to go was too far.

Hence Tibet. An hour into the climb the sun glared directly above us in the narrow gap of sky not shuttered by mountains, so we stopped to reapply sunscreen. I smeared more on my face and Mel daubed some on her lower back, not so much to block it from UV radiation, for it was already shielded by her shirt, but to moisturize her skin, which was flaking away in bright red bits. The day before sneaking across the checkpoint, while bending over to sort and pack gear, Mel's T-shirt had slipped up and exposed her lower back to the high-altitude glare, which fried the skin a few shades angrier than scarlet. She didn't complain—Mel rarely did except to exalt her suffering in satire, a form of stoicism I admired and occasionally found insufferable—but I could tell she'd been riding stiffly to avoid twisting her seared torso. No easy task on a road paved in potholes.

After moisturizing, Mel sighed in the quietly determined way that meant she was ready to get back on the road. I wasn't. "Do you hear that? Is it a bird? Or maybe the boys?" I ventured, hoping to distract her into resting a little longer. A large part of why I love biking is how blissful it is to stop. "Hey, are you hungry?"

Of course she was; we always were. Though we'd packed all the basic provisions we'd need for the next month, from oatmeal to instant noodles, our appetites far exceeded the carrying capacity of our panniers. The day before, we'd even contemplated eating the goat handed to Florian by a passing Chinese motorcyclist, who probably thought we all looked undernourished. Cradling the bendy creature in his arms, Florian, a gentle mathematician incapable of slaying more than differential equations, had gazed at the rest of us questioningly. Mattias licked his chapped lips. Ben nodded in a kind of drooling trance. The goat, in a shrewd move, levelled its sweet cudgel face and Elvis Presley bangs at Mel and mewled adorably.

"That's it, boys, give him back!" she said in a way that hinted she'd eat Ben and Mattias before allowing the goat to come to harm. A vegetarian who melted for anything four-legged and fuzzy, Mel had recently squandered half an hour of precious videotape filming baby goats frolicking at a gas station. Florian looked visibly relieved at Mel's suggestion and carefully handed the goat back to the Chinese motorcyclist, who stuffed it in a burlap saddlebag and rode away. So now we sat on the shoulder of Highway 219 and ate stale cookies instead.

"Typical," Mel muttered after taking a bite.

"What?"

"Not a *single* chocolate chip."

Mel was frowning at her half-eaten cookie, whose glossy packaging had promised a dense cosmos of chocolate chips. I'd scarfed my own cookies so fast I'd barely noticed the lack. This wasn't the first time we'd been duped by misleading advertisements in China. In the two months we'd biked through Xinjiang before steering for Tibet, Mel and I had purchased popsicles that claimed to be strawberry, watermelon, fruit punch, and chocolate in flavour, only to discover they all contained, without exception, the same tasteless puck of brown ice flecked with red beans. Beans! Who puts legumes in popsicles? And why on earth did we keep buying them?

Perhaps out of the same reckless optimism that saw us sneak illegally into a land almost as oxygen-starved as Mars. Or the stubborn faith of pilgrims who repeat the same mantra, convinced it will eventually take them to a different place. Back on the bike, I pretended that the wheels didn't travel the world's surface so much as unspool it, and if I stopped pedalling for even a second it would all fade away. The mineral glitter of the mountains and the cloud-shot indigo sky and this road like a parade of detours was all a dream sustained only in motion.

Three hours and as many false summits later, I knew we'd reached the top when Mel, ahead on the road, threw her bike down and started turning cartwheels. I was so light-headed and giddy I seemed to be cartwheeling while standing still. It was one of those rare moments in life when you measurably accelerate into a new version of yourself, become who you are by leaps and bounds. That I'd pedalled to an altitude I'd only previously visited in airplanes, and that I could still breathe, was a revelation, like discovering an extra lung or the ability to see in ultraviolet. I'd always hoped we'd make it to the Tibetan Plateau, still technically a few passes away, each higher than the last, but now, for the first time, I *believed* it.

Mel and I celebrated over hot chocolate as we waited for Ben, Mattias, and Florian to catch up. The drink was a product of China, meaning its cocoa content was limited to the design on the packaging, but context alone provided ample flavour: anything tastes delicious when you're high in the Himalaya with your best friend, utterly wiped but eager for more—more wending road, rough peaks, deep and indivisible sky. More of anything that goes on and on.

The descent rather achingly met that definition. As we sped down the pass, every little bump and divot and pebble on the road blurred together into a pavement of pure concussion. Such is the price you pay to reach forbidden Tibet: pain in the legs, in the butt, and in the brain, which can't conceive a coherent thought because all it knows is the jackhammer jolting of the body and bike to which it is connected. I'd known that climbing the pass would be tough, but I'd never guessed that coasting down it would be tougher. By the time we pitched the tent in the glacial rubble of a valley I had a throbbing headache. Bolts of pain arched between my brows. I collapsed in my sleeping bag, sure I couldn't go on.

Yet the next day I woke up eager to meet the oncoming rush of road. Maybe it was the resuscitating power of instant oatmeal mixed with peanut butter. Maybe it was the Nescafé 3 in 1 with which Mel and I washed down that glutinous mash. Either way, I set off each morning feeling strangely convinced that I was on the verge of some grand discovery, despite travelling a bygone trading route on the world's most populous continent. This was hardly terra incognita, but it sure felt that way, especially when Mel and I biked up a seventeen thousand–foot pass a week later, the highest and least oxygenated stretch of our Silk Road yet, and barely came down on the far side.

Suddenly the land was spread wide as wings, sloping here and there into mountains. No trees, no greenery, no colour anywhere, really, except for turquoise salt lakes glimmering in the distance like puddles of sky. The horizon was more hesitation than a hard edge, and every so often it spat out a dust tornado that would skim across the road in eerie silence just a few metres ahead of us, the flue curved into a question mark missing its point. Where on this spinning world was I?

A place where mountaineers find fossil seashells on summits, where the flattest plains are higher than the tallest peaks in the contiguous United States, where the wind carries the tang of salt and every horizon has a distinctly oceanic fetch. Welcome to the Tibetan Plateau, the loftiest sweep of land on the planet, a kind of perfect compromise between heaven and earth.

Mel and I happened to visit the plateau during a lull in its anguished modern history. It was the summer of 2006, a few years before the violent crackdowns against Tibetans in the lead-up to the Beijing Olympics, before the forced resettlement of nomads into bleak subdivisions, before self-immolations by monks and nuns became regular news. The Chinese authorities simply couldn't be bothered with a bunch of wild-haired cyclists, though Mel and I knew this could change at any moment, which is why we were spooked when a military truck roared up behind us and dispatched a pair of soldiers.

The men wore tinted sunglasses, black boots, and jungle-green army fatigues, which seemed odd camouflage in a landscape the opposite of lush. When they grabbed our bikes we feared the worst, but they only wanted to ride them. The soldiers took turns wobbling down the road, snapping photos of each other with their cellphones, breathing like they were blowing out birthday candles. After a few laps they gave us back the bikes and waved goodbye, apparently not caring that we were about to trespass onto the Aksai Chin.

This particular stretch of salt and wind, nearly uninhabited and widely dismissed as a wasteland, is one of the most contested territories in Asia. Tibetan by cultural heritage, Indian by treaty claim, and Chinese by possession, the Aksai Chin is caught in this territorial tug-of-war owing to its strategic location between nations. It all began when China furtively built a road across it in 1957, the very dirt track we were on, roping like a slow-burning fuse for more than a thousand miles over the emptiest edge of the plateau. India only clued in to Highway 219's existence half a decade later, and their discovery detonated a war over the borderland. Hundreds of Indian and Chinese soldiers died by grenades, machine gun, and mortar fire to claim a place Jawaharlal Nehru, then the prime minister of India, described as so barren "not a blade of grass grows." Even today most of the Himalayan frontier between India and China falls in disputed territory, with big chunks of the border still a blur, as though someone had smudged the ink on the map before the labels and lines had a chance to dry.

Not that the Chinese road atlas Mel and I carried revealed any of this, its pages greasy and almost see-through from being handled by sunscreened fingers. What the maps made clear, though, with a lasso of bold strokes, was that the Tibetan Plateau was unambiguously Chinese. This ownership contradicted the map the Germans carried, which diplomatically marked the Aksai Chin with dotted lines.

We're so used to thinking of nations as self-evident, maps as trusted authorities, the boundaries veining them blue-blooded and sure. In places like Tibet, though, the land itself gives those lines the slip. Borders might go bump in the night because they're reinforced by guardrails, but also because they exist in only the most suggestive, ghost-like ways. At least that's how I sensed them on the Aksai Chin—as a kind of haunting presence on horizons otherwise fenceless and patrolled only by wind. What if borders at their most basic are just desires written onto lands and lives, trying to foist permanence on the fact of flux?

A sand tornado spun past me, trailing its skirts of dust. I inhaled the country and kept pedalling. Then I realized the vortex came from the Chinese military convoy speeding up behind us. We scooted over to let the vehicles pass, dozens and dozens of black jeeps in a long litany of exhaust. But even as I felt unnerved by the sight of soldiers patrolling the Aksai Chin, what chilled me even more was how I suddenly saw myself in them. "Longing on a large scale," says novelist Don DeLillo, "is what makes history." And longing on a smaller scale is what sends explorers into the unknown, where the first thing they do, typically, is draw a map.

Admittedly, I'd spend more time musing about borders and explorers once back home and better able to breathe. For the time being I was preoccupied with staying upright on my bike. Some days the road was barely there, a faint scar in the sand or a spill of rocks indistinguishable from the rest of a mountain's rubble. At one point it disappeared entirely beneath a stream. Glacial meltwater sluiced between my toes as I dragged my bike through it, and I couldn't feel my feet for hours. Also reliably freezing were the headwinds, bitter and constant, as if hidden deep in the Himalaya was the world's core of wind, sculpting the planet as it streamed off glacial ice.

It didn't help that our loaded bikes were effectively bulky sails, heaped as they were with tents, sleeping bags, spare parts, tools, and food—all the instant noodles, peanut butter, and misleadingly packaged snacks we needed to survive, barely. Even if grocery stores and restaurants had been commonplace in western Tibet, Mel and I had no cash left to buy fresh rations, at least none accessible. We'd stashed our last hundred-dollar bill in the hollow metal tubing of my handlebar, making a piggy bank of my bike, but after months on the road the money was stuck inside, effectively welded to the aluminum.

Caloric relief occasionally arrived in the form of meals shared with us by Tibetans. One moment we'd be alone on the plateau, and the next we'd be surrounded by nomads who materialized from the mountains. A waft of smoke and sheep's wool on the wind and there they were, men and women and children with burnished copper faces and chapped red cheeks and thick ropes of licorice hair. The men wore fancy if tattered dinner jackets and jaunty felt hats. The women's necks were slung with bright chunks of amber and turquoise and coral. When we were lucky, they invited us back to their tents for tsampa, a gruel of roasted barley, and yak butter tea, a brew that congealed as we sipped it.

"Tastes like . . . animal?" mused Mattias, his upper lip shiny with lard. He slurped again at the thickening slurry and nodded. "Tastes like yak."

The canvas tent breathed in and out. An elderly man beamed at us as he absently rubbed a string of wooden beads, his calluses like dark coins in his palms. We sat on hard wooden benches and sipped our tea, our eyes slowly adjusting to the darkness and sting of dung smoke. It was a simple home with a thousand gleaming surfaces: porcelain-looking cups and bowls, tins whose Chinese labels I couldn't decipher, pots and kettles, a clock that didn't work, its thin hands unsteady in the wind that shook the fabric walls. I had no idea how the family moved all of this from camp to camp, or how often they moved camps in general, or a million other details I longed to learn but lacked the words to ask about, so I just smiled dumbly and scooped buttery clumps of tsampa into my mouth with my fingers.

Across the tent, tacked to its supportive beams, a glossy poster caught my eye. It featured juicy-looking burgers, golden french fries, bowls of cherries and oranges and ice cream, and foamy milkshakes all spread on a red-and-white picnic blanket in a lush forest next to a waterfall. We'd seen similar posters all across western China, whether in Han restaurants in Kashgar, Muslim mud-brick huts in Xinjiang, or Buddhist camps in Tibet. They fascinated me not just for the torturously improbable feast they portrayed—food that was the stuff of fantasy, unavailable for thousands of miles—but for the odd familiarity of the scene. For all I could tell, the posters showcased woodsy rural Ontario, where my own bedroom walls had been tacked with posters of mountains and deserts, of horizons picked clean by wind. We were longing right past each other.

After the meal, Mattias brought out his German edition of *Seven Years in Tibet*, which is about an Austrian mountaineer's escape into Tibet from a prisoner of war camp in India. He flipped to the middle of the book and showed our hosts a black-and-white photograph of the young, grinning, bare-armed Tenzin Gyatso, His Holiness the fourteenth Dalai Lama, and winner of the Nobel Peace Prize for his nonviolent opposition to China's occupation of Tibet. The Tibetans craned close to catch a glimpse of him.

Under Chinese rule, it was illegal to possess a photo of the exiled spiritual and temporal ruler of Tibet—a place more accurately named "Bod," as Tibetans have referred to their homeland throughout recorded history. From the seventh to ninth centuries, the glory days of the Tibetan Empire, Bod stretched into parts of modern India, Pakistan, Afghanistan, Tajikistan, Kyrgyzstan, and China. Roughly six centuries later, when Marco Polo followed the Silk Road onto the plateau, he arrived in the much smaller territory of what he called "Thebeth"—less the name of a coherent nation-state, as the modern concept goes, and more a geographic description from an old Turkic word for "heights." By then Tibet was under Mongolian rule, though Polo reported that the inhabitants of that "desolated country" refused to use Kublai Khan's paper money, preferring instead their usual currency of salt.

Over the next few centuries, Tibet was variously administered by dynastic China, under attack by the British Empire, or enjoying a rare lull of peace and self-rule. The latter ended in 1950, when the People's Republic of China invaded the Buddhist country and eventually forced the Dalai Lama to sign over Tibet's sovereignty. Eight years later, when the Chinese violently suppressed an uprising in Lhasa, the Dalai Lama fled for his life into India and tens of thousands of Tibetans followed him. Many hundreds of thousands more have fled since, for the country they once knew was "liberated" by the Chinese in 1959, meaning the former government of Tibet was declared illegal and the once-independent nation was forcibly downgraded to the not-so-autonomous region of Xizang, which in Mandarin means, rather tellingly, "west treasure vault." Ever since, the plateau's vast reserves of copper, lithium, gold, and silver have funded China's economic growth, and Tibet's borders-turned-regional-boundaries have hosted checkpoints that restrict mobility— not of foreigners, as we experienced firsthand, but of locals. No wonder Tibetans are reputed to be such students of impermanence. As empires flourished and fell at their feet, as their own frontiers expanded and shrank and turned hard against them, daily life on the plateau had proved the illusoriness of any firm place to stand.

Mattias ripped the photograph out of the book and offered it to the old man, who touched it to his forehead and then folded it into his cloak for safekeeping, taking care not to crease His Holiness' face. We thanked the family and got up to leave, wiping our buttery hands on our bike shorts. Back on the road, the tent looked so small when I glanced behind me, a tiny white envelope stamped with prayer flags. Yellow, green, red, white, and blue, with each colour signifying an element and state of mind, and each flag inscribed with sutras on desire and suffering, compassion and flux—the kind of writing that goes further the more it fades away.

Maybe Pangong Tso, a lake that spills across Tibet into northern India, represents the most honest kind of borderland: a frontier defined seasonally by changes of state, solid to liquid to air. The water was so vast and turquoise it looked tropical, like a remnant of the ancient Tethys Ocean, whose warm blue waters were swallowed beneath the Indian subcontinent when it slammed into Eurasia fifty million years ago, crumpling the sea floor into the Tibetan Plateau. At nearly fourteen thousand feet in elevation, the lake's inviting appearance belied a more frigid reality, but we hadn't bathed in weeks.

Mel, the Germans, and I had ditched our bikes and immediately dove in the water. I bobbed on my back, the brisk water soothing my saddle sores. Luckily they weren't as gory as when Mel and I pedalled across America, only to learn, thousands of miles too late, that wearing underwear beneath padded bike shorts is a major faux pas. Here the water gently made off with all my weight, the pull of earth and sky precisely equal and opposite. I tried to make out Ladakh, the region in northern India known as "Little Tibet," but the lake was too long to see the far side. What I could see was Ben, still lingering anxiously on the shore. Finally he waded in, frowned at the water around his legs, and stormed out again, claiming he saw an oily film on its surface.

"Ben, it's you," we tried to reason with him. "It's your filth, your sunscreen."

But he was already towelling off in a fury. An old collarbone injury of his had flared up again on Tibet's terrible potholes, and his mood wasn't improved by the fact that this should've been the easiest stretch of the trip—roughly a hundred miles of gentle, flowing downhill, according to the map, but what the topographical contour lines failed to convey was the road's sandy, half-sunken texture. Adding insult to injury for Ben were the large piles of gravel deposited every few feet along Highway 219, hinting that any day now a maintenance crew would firm up the sand, fill the potholes, and smooth the washboard ruts. But as the weeks went by, so did the road maintenance trucks, one, two, sometimes three a day, crammed full of Chinese workers in yellow safety vests. "Do your job!" Ben raged as they passed. "Fix the road!" But the Chinese workers, confusing his yelling and flailing for a friendly greeting, smiled and waved merrily as they sped off to repair some other stretch of road.

The plateau, Ben told us in not so many words, was hardly his idea of Shangri-La. This was another label Tibet had unwittingly earned over the years, mostly thanks to James Hilton's 1933 novel *Lost Horizon*. Hilton's book and the blockbuster film based on it depicted a lush, paradisiacal Himalayan aerie hidden in Tibet, falsely fixing the place in popular imagination as a kind of pre-lapsarian refuge, a place of mystical innocence and immortality. The real Tibetan Plateau, or at least the western corner we'd biked so far, was, by contrast, "beyond doubt among the world's bleakest stretches." At least that's how Nehru described the Aksai Chin, that so-called wasteland for which he nevertheless waged war.

Truth be told, the western plateau's bleakness was so other-worldly, so breathtaking, that I could understand why everyone wanted to claim it for themselves. Tibet is often romantically evoked as the roof of the world, as if the plateau served as some kind of elaborate shelter, but it was raw exposure that I craved and found there. What the plateau truly presents is not refuge, but a new frame of reference: from those dizzying heights, you can glimpse the *real* roof of our world, that faint swaddling of oxygen and nitrogen that holds us back from the heavens, or the heavens back from us. A thin blue rim, barely sixty miles thick that buffers all life on Earth from the bottomless void of space.

Sixty miles was also the kind of distance I could bike in a day, if only I could bike straight up into the sky, but pedalling across Tibet was gruelling enough. Which was part of the plateau's charm to everyone but Ben, who wore headphones much of the time now and biked for hours with the music blaring, wilfully deaf to wind and the clatter of wheels on that gorgeous, gut-shaking road. Mattias had generously loaned Ben his iPod featuring a limited selection of music on repeat, including the *Baywatch* theme song and Avril Lavigne's "Sk8ter Boi." I could imagine few things more punitive than such a soundtrack, but judging by the distant, blissed-out look in Ben's eyes, the music succeeded in sweeping him away, elsewhere, home, the late 1990s, anywhere but there and then in Tibet, the only place in the universe I wanted to be.

A month after crossing the checkpoint in Kudi, the four of us reached the small city of Ali, a relative metropolis for western Tibet and the end of Ben's Silk Road. He was tired of being per-petually tired, and tired, no doubt, of the fact that the rest of us were loving every torturous minute. With only a week or two left on his Chinese visa anyway, he hitched a transport truck from Ali to Lhasa, boarded a plane to Beijing, and flew home to Manitoba. Mel and I gave him a package of dehydrated yak penis as a depar-ture gift, which Canadian customs, we were sad to learn, immedi-ately confiscated.

According to other cyclists who'd snuck across western Tibet, it was possible to turn yourself in to the police in Ali and be granted temporary legal status, which would make it easier to navigate the greater density of checkpoints on the way to Lhasa. So Mel and I took deep breaths and surrendered our documents at the police station. After filling out a bunch of forms we couldn't read and looking suitably repentant, the officers grudgingly returned our passports, now containing slim vouchers titled "Aliens' Travel Permits." "Congrats, Kate," Mel said as we exited the station. "Legally recognized as a Martian at last!"

I did feel strangely at home on the Tibetan Plateau, a sense of deep arrival that was almost disordering. I don't mean to claim a cheap affiliation with a culture or complicated history that, at the time, I barely knew, and still am only learning. Instead what I felt was an affinity for the land itself, the stark contours and harsh tectonics of the plateau, this territory of uplift and change. As Mel and I biked out of Ali, I couldn't stop thinking about the "pale blue dot" photograph taken by Voyager I before the spacecraft sped out of the solar system forever. The image revealed our home as a tiny speck of blue in the darkness of deep space, "a mote of dust suspended in a sunbeam," in the words of astronomer Carl Sagan. Although the Voyager's instruments dutifully recorded the size of particles in Saturn's rings, among other tasks on their strictly scientific mission, that casual snapshot offered something far more rare and significant than reams of data: a fundamental change in perspective. Wasn't that the most meaningful outcome of any kind of exploration? To reveal the old world—and ourselves—anew?

The Tibetan Plateau offered a similarly cosmic reality check. There I was, little more than a mote of dust myself after a month without showering, biking slowly among summits that had once been sea floor. It was like being on the moon or Mars, only better: I could breathe, laugh out loud, feel the wind on my face. I didn't have to report to Mission Control or speak through radio static. The only time I felt briefly nostalgic for the protective buffer of a spacesuit was when my bike flatted a tire as I was speeding down a hill, throwing me off balance and into the dirt.

"Nice air, Kato!" Mel said after she made sure I wasn't hurt. I picked bits of gravel from my bare palms, regretting the fact that I hadn't been wearing gloves. After absorbing sweat, dust, and sunscreen for months on end they were effectively casts, so I'd opted to go without them despite the high-altitude chill of August. Mel helped me patch the inner tube, already a quilt of other patches, and she steadied the wheel while I inflated the tire with a portable bicycle pump that squeaked with dust. Not until I was back on my bike did I notice collateral damage from the crash: I'd sliced open my down jacket on a rock and now tiny white feathers wafted off me, as though I were moulting.

Clouds slid down the slopes of the valley, now ambered in a slow, tilting light. The air was cold and spiceless, and the wind moved like something alive. Despite the duct tape I slapped over the tear in my jacket the odd bit of plumage still escaped, but I didn't mind. The plateau could have my feathers and sweat and even the skin off my palms—whatever it took to be here, to earn this intimacy with immensity. If to be an explorer I must draw a map, I remember thinking, let it be this: How the sky shifted and darkened over the plateau that night, and the sun gave a last golden glance through the clouds. How the mountains shone like bits of fallen moon all around me, glowed for a moment and were gone.

3.

NATURAL HISTORY

England and New England

A few weeks later Mel and I flew back to Canada, dust sewn in our sleeves like the jewels Polo reportedly smuggled home in the seams of his clothes. I watched China shrink in scale and tucked into a microwaved airline meal, food only marginally more flavourful than instant noodles. After biking four thousand kilometres in four months, it was almost a relief to sit still for a while, to speed along without pushing pedals. That didn't last long. Approximately one in-flight movie later I felt restless again, and so did Mel. Before the plane touched down in Toronto, we vowed to someday ride the rest of the Silk Road, namely the rather prodigious gap between Europe and Asia.

I didn't bother unpacking my bike before checking it onto another plane to England, to begin graduate school at Oxford, where some new friends and I kickstarted our studies with a cycling trip. We figured we had just enough time before classes began to ride to Stratford, the birthplace of Shakespeare, where Patrick Stewart, otherwise known as Captain Jean-Luc Picard of *Star Trek*, was starring as Prospero in *The Tempest*.

I kept expecting to see the Tibetan Plateau when I woke up in my ragged sleeping bag and zipped open the tent, only to behold rolling green England instead, terrain more like the pastoral posters I'd seen in China. For two days, Dominique, Kim, Jamie, and I biked past thatched medieval villages and castle-like estates with long prim lawns that doubled as runways, which we knew for certain because we saw a plane land on one of them. We foraged so many blackberries from hedges that our lips and tongues were stained a royal purple by the time we arrived in Stratford, only to learn the play was sold out. In hopes of getting rush tickets the next morning, we pitched our tents at the front of the line on the terrace of the Royal Shakespeare Theatre, using the weight of our bikes instead of tent pegs to pull the structures taut. The crucial flourish for our campsite was a hand-scribbled cardboard sign that read "Patrick Stewart's Biggest Fans."

In truth, none of us were hardcore Trekkies. I'd enjoyed reruns of *The Next Generation* from time to time, but my brothers and I had preferred the *Voyager* series, in which Captain Kathryn Janeway not only boldly went where none had gone before but found herself stranded there for decades. Shakespeare was the main draw for Dom, a free-spirited literature major from Quebec, and the bike trip itself rather than its destination appealed to Jamie, a lawyer with flaming hair who'd been brought up in Yukon bush camps. As for Kim, an aspiring family doctor from Ontario with a bedside genius for comedy, she was game for anything that made a good story. We'd all won a Rhodes scholarship, and flew to London on the same red-eye flight. When Kim, Dom, and Jamie were game to begin grad school with a bike trip, despite their jet lag, I knew we'd all get along.

At dawn we woke up to discover three nervy people standing ahead of our tents in line. Fortunately there were enough rush tickets to go around. After brushing our teeth and washing our faces in the theatre bathroom, we found our seats for the matinee performance. The lights dimmed, an eerie soundtrack struck up, and Shakespeare's words transformed the theatre into a bleak Arctic island where it never stopped snowing, where sanity and madness were just a thin pane of ice apart, and where Jamie and I held hands in the darkness of the polar night.

"O brave new world!" exulted Miranda. "That has such people in't!"

"Tis new to thee," her father, Prospero, wryly clarified.

A mutual acquaintance had put Jamie and me in touch over email when he realized we were both heading to Oxford in the fall. As I was cycling the Silk Road and sneaking into Tibet, Jamie was studying Arabic in Egypt and motorcycling into Syria. I read his travel blog in the smoky Chinese Internet cafés where I posted my own missives from the road, which Jamie read in turn, and we recognized in each other a common longing and lament—for the faraway and wild, for the loss of both from the world. "It is like having, as Pascal said, a God-shaped hole in your heart, but the hole is filled by empty space, silence, and nothingness," he wrote from Egypt, articulating the nameless longing that I knew so well, and that only mountains and deserts seemed to satisfy.

That September we each travelled to Ottawa for Sailing Weekend, when the latest batch of Canadian Rhodes scholars gathers to be wined and dined and toured around Parliament before launching across the Atlantic, though sadly no longer by boat. I feared I wouldn't be going to England at all. Between returning from the Silk Road and heading to Ottawa, I'd barely had time to mail off my passport, which meant my student visa hadn't been processed yet. Fortunately Arthur Kroeger, a kindly old Rhodes scholar and legend in the Canadian civil service, reassured me he'd sort things out through his connections at the British embassy. He mentioned that another scholar was in the same situation.

Of course it was Jamie, just back from Egypt. His face was extremely pale, his red hair a lit torch. He spoke with a deep-keeled intensity and also a sense of his own smoothness, just the type you'd peg to win the World Universities Debating Championship—except for what he did afterwards. The tournament was in Malaysia that year, and as the debating began a tsunami struck southern Thailand. Instead of flying back to Canada to finish his law degree after his win, Jamie travelled to the devastated Thai coast and spent a month helping to clean up the wreckage. He was fascinated by wreckage; he had an anthropological obsession with it. We kissed the first night we met.

That was in Ottawa, not long after we received our student visas. When a bus delivered us a few days later among Oxford's dreaming spires, I remember marvelling over how a space cadet from small-town Ontario had ended up in this fairy tale. Not just the budding romance with Jamie, but the scholarship itself. Surely it was a mistake, some confusion on the part of the selection committee, but I planned to run with it as far and wide as I could. Though I'd originally intended to study science at Oxford—part of my overarching mission to become an astronaut and launch to Mars—I switched at the last minute into a master's degree on the history of science. I'd be doing science for the rest of my life, I reasoned; why squander two years at Oxford in a laboratory? Especially when all laboratories are exactly alike: sterile, impersonal, replicable by necessity. The Bodleian Library, by contrast, has no parallel.

Loosely supervising my studies was Professor Pietro Corsi, an operatic Italian in his late fifties who got all flushed and passionate about Darwin's seven-year obsession with barnacles. Corsi's specialty was the history of evolutionary theory, and his lecturing style was quaintly non-linear. "I had dinner last night with a very senior biochemist, eighty-four years of age, who talked about the history of science like he was a positivist living in the 1870s!" Corsi exclaimed, shaking his head ruefully. I exchanged baffled looks with the other grad students. "You see," he continued, waving his tweed-clad arms, "a belief in logic, in rational progress toward truth, is a seductive thing for a scientist. I mean, I have no doubt at all that he is wrong, poor man, but you cannot kill him!" Then he levelled a dark, knowing look around the room. "This Oxford is a curious place. Everything here is forbidden, and that is why everything is possible . . ."

Most aspects of Oxford—from the twisting cobbled streets to Corsi's lectures—encouraged digression, which is, after all, just a sideways method for stumbling on connection. Such as between the philosophy of science and poetry, if one were to go by Emily Dickinson's definition of the latter as whatever makes you feel as if the top of your head has been physically taken off. I tucked into bed one evening intending to put myself to sleep by skimming a few chapters of Thomas Kuhn's *The Structure of Scientific Revolutions*, which our class was discussing the next day. Instead I stayed up all night reading with demented avidity to the final page, my empirical understanding of the world undone by Kuhn's argument that scientific theories are in essence evolutionarily selected stories, that is fictions that best fit the available facts—until the discovery of new facts forces a paradigm shift to a different and better fiction. More than that, he argues that scientists who embrace a new paradigm at an early stage—before sufficient evidence has been amassed to trigger a scientific revolution—do so not out of a sober consideration of the available facts, or at least not *only* that, but also with a subjective, irrational, from-the-gut leap of faith. Reading Kuhn and various other philosophers of science was like peering into the skull of a scientist, one of those Spock-like arbiters of imperishable truths, only to discover a raving mystic inside. I was rather partial to mystics, especially the writer Annie Dillard, but I never dreamed their motives and methods didn't differ drastically from those of scientists, though in retrospect Dillard's books should've prepared me for this. "What is the difference between a cathedral and a physics lab?" she asks in one of them. "Are they not both saying: Hello?" At dawn I put Kuhn's book down and tapped the top of my head to make sure it was still there.

Another class at Oxford revealed the connections, more depressingly, between war and science, the two ends of Galileo Galilei's telescope. The Italian professor of mathematics was awarded tenure at the University of Padua for his vast improvements to the design of a spyglass for military purposes. Of course Galileo himself used the device to more peaceably spy on the heavens and in the process observed craters marring the moon, spots blemishing the sun, other moons orbiting Jupiter, and the fact that Venus has phases—observations that collectively threw the static universe into motion. All I saw, squinting through an exact replica of Galileo's spyglass while standing on Broad Street at noon, wearing the clammy purple lab gloves our professor had provided to protect the gold-flecked leather tube from fingerprints, was the sign for the popular King's Arms pub, its letters blurred and distorted by the glass—not so different, really, than how they looked to students exiting the pub.

But the best part about studying the history of science? I suddenly had to do for homework what I normally did for fun: read expedition journals, such as Charles Darwin's from his voyage on the *Beagle*.

Though I'd known about Darwin since high school, I'd never read his diaries. They revealed that when he set sail for South America, at twenty-two, Darwin was little more than the ne'er-do-well son in a well-to-do British family. After failing to finish medical school (he couldn't stomach the sight of blood) and failing to become a countryside parson (he was more concerned with collecting beetles than saving souls), Darwin begged his exasperated father to let him join the *Beagle* expedition—not as a naturalist, but as a gentleman companion to Captain Robert FitzRoy, who feared going mad if deprived of dignified society for years on end. Darwin even had to fund his own way, though he was subsidized by his affluent family.

Once the *Beagle* set sail, Darwin was in his element—except for his chronic seasickness. Judging from his diaries, the young naturalist seemed driven by the same restless, rangy impulses I recognized as my core. He exulted in the "strife of the unloosed elements" and the "inexpressible charm" of living in the open air. Wending along the coast of South America, and spending long stints on shore, Darwin was so staggered by what he saw—oceans blushing with chameleon octopi in Cape Verde, skies snowing butterflies in Patagonia—that he confessed at times he was scarcely able to walk. In the closing lines of his journals, Darwin urged aspiring young explorers to take all chances and start on a long voyage— by land if possible, he recommended generously, hoping others might avoid the queasiness he was never able to shake.

I fell in love with this wide-eyed, seasick wanderer through his diaries, but upon further reading fell out again. After six years abroad, Darwin returned to England and in short order secured himself a wife, settled in a country cottage, and never travelled anywhere again. To be fair, it was during this transition from restless to rooted that Darwin elaborated the theory of evolution by natural selection. Fathering ten children, including one who died prematurely, and mysteriously poor health eventually anchored him in England. But what crushed me about Darwin was not that his long voyage ended. From reading Henry David Thoreau I knew you could travel widely from a cabin in Concord, and I hoped to someday do the same from a cabin in Atlin. Far more disturbing than Darwin staying home was his withdrawal from wonder.

In his frank, confiding autobiography, Darwin describes how he turned into "a kind of machine for grinding general laws out of large collections of facts." As he honed his taxonomic identification skills studying barnacles, pigeons, and other specimens from the *Beagle*, Darwin noticed himself becoming increasingly numb to music, poetry, and nature. "I retain some taste for fine scenery," he confessed, "but it does not cause me the exquisite delight which it formerly did." Science became the elder Darwin's exclusive passion, but that term usually connotes some measure of enjoyment, and what he seemed gripped with instead was a cold mania for sorting facts into theoretical frameworks. Meanwhile he lamented the withering of his more whimsical, imaginative sensibilities as a "loss of happiness."

I couldn't believe the younger and older Darwins comprised the same person. His transmutation from madcap wanderer to morose scientist gave me chills, but I never imagined it could happen to me. Hadn't I felt like Ralph Waldo Emerson all my life, crossing a nondescript field on a cloudy day, "glad to the brink of fear"? That's how I felt walking out of my dorm at Hertford's graduate residence, crossing a bridge over the River Thames, meandering along paths shaded with towering oaks in Christ Church Meadow, and then down a narrow cobbled street and up a creaking flight of stairs to where Jamie lived.

Though we were fairly inseparable that first term, Jamie still wrote me letters, beautiful handwritten meditations I'd find in my pigeon hole, or "pidge," as mailboxes in the porter's lodge of Oxford colleges are called. "Real life is mental life, spiritual life," he wrote in one such missive. "Wagering your soul is a real wager. As Benedict Allen said of travel and exploration"—we'd recently gone to a campus lecture by the British adventurer—"it's not about making your mark on a place, but about letting it make its mark on you."

It wasn't hard to imagine such missives being delivered by birds, the duck and swerve of words among Oxford's mist-sleeved towers and leering gargoyles. Everything was forbidden, therefore everything was possible. "Keep off the Grass" read the signs posted on prim green lawns in the imposing stone quadrangles across campus, but manicured terrain held little appeal anyway. Instead I climbed over the stone-and-metal gate at Magdalen College into the deer park where C. S. Lewis used to stroll while dreaming up Narnia. Or I went running along the Thames, arriving back at my dorm with just enough time to shower and put on the only dress I owned: an elegant black gown studded with faint stars, its synthetic fabric so immune to wrinkles I could stuff it in a backpack, forget about it for days, then wear it to a formal ball. In Oxford, it seemed, they took place every other week.

As Jamie and I walked home from one such event, the streets at midnight were full of students in fancy dress trying not to trip on cobblestones, and the champagne glow of the sodium lamplights was just dim enough to let some stars through. What I felt then came to define the radiant, widening two years I'd spend at Oxford: the delicious sensation of getting away with something, like I'd given real life the slip.

~

One weekend in early winter, when the English rain was relentless, Jamie swept me off to Cinque Terre, Italy, a destination he chose because it was sunny, abundant in cheap red wine and pesto, and also where Percy Bysshe Shelley drowned. We swam in the spit-warm turquoise waters of the Mediterranean and then read the doomed poet's words to each other as we sunned dry on rocks. Jamie was somewhat obsessed with Shelley, the renegade Romantic poet who was expelled from Oxford for his atheism, only to later be apotheosized in a larger-than-life marble nude in one of the colleges. Many of Oxford's most famous alumni never actually completed their degrees, from Shelley to former US president Bill Clinton, and Jamie openly aspired to follow in their footsteps. A lawyer turned Development Studies student, he skipped class often, wrote letters to me instead of term papers, and airily proclaimed that he "didn't believe in development." For his thesis he planned to write an aesthetic refutation of the whole enterprise, fairly certain it would rile his supervisors and get him ousted from the establishment.

While I didn't plan to fail out if I could help it, I knew that nothing I did at Oxford really mattered in the scope of my extraplanetary ambitions: a Ph.D. admissions committee at MIT, where I hoped to study microbiology in extreme environments, wouldn't care about my results in a humanities degree. I'd made my way to England by being good at school, hence the scholarship to Oxford, but once I got there I almost learned not to care about it, or rather to care for the right reasons: not as a means to a Martian end, or success as sanctioned by others, but as an opportunity to think, dream, stray out of bounds. A venue, in other words, for exploration.

What would you study if there was no such thing as making the grade? In my case, I obsessed over the reports of early Himalayan explorers and scrutinized centuries-old survey charts in the Bodleian Library, trying to glean the logic behind the lines Mel and I had seen (and snuck across) on the Tibetan Plateau. On every map I examined, a huge bleed of white to the west of the Aksai Chin caught my eye, in part because it reminded me of the Juneau Icefield. This particular enormity of slow-flowing ice, I learned, was the Siachen Glacier, one of the last unexplored gaps on the map until the early twentieth century, when the redoubtable Mrs. Fanny Bullock Workman hitched up her tweed petticoat and hiked onto its base.

In her mittened hand she gripped a sign declaring—not asking, thank you very much—"Votes for Women." Plodding breathlessly a few steps behind her was Dr. Hunter Workman, and behind him were a dozen porters hired to ferry their gear. The Workmans were wealthy amateur naturalists from America, a husband-and-wife team. After a doctor prescribed fresh air and foreign travel as a cure for Hunter's chronic lassitude, the two of them launched on cycling journeys through Spain, India, Burma, Ceylon, Java, and parts of Africa. When they ran out of roads, they began trekking in what was then Baltistan in British India, which centuries before was a part of Tibet, and today is contested Kashmir. From there they crossed the Karakoram Pass and followed a southern route of the Silk Road into terra incognita: the Silver Throne plateau of the Siachen Glacier, where Hunter snapped an iconic photo of Fanny in a tweed dress and ribboned hat, championing suffrage at 21,000 feet.

Siachen roughly translates from Balti as "the place of wild

roses" and is named for the hardy flowers that take root in its glacial till. According to Fanny's 1917 book, *Two Summers in the Ice-Wilds of Eastern Karakoram*, she preferred to call the glacier "the Rose," pleased by the incongruity of this dainty label applied to a violence of rock and ice. Fanny claimed she wanted to go there for strictly scientific reasons, to survey the glacier and triangulate all its important peaks, but this smacked to me of logic appended to pure longing, and I would know. All I'd really wanted to do on the Juneau Icefield was wander around, see the world from a different point of view, though if anyone asked, I'd come to study the geophysics of glacial flow. I was smitten with wildness, and only incidentally with science, and I suspected the same was true for Fanny.

Not that she wasn't a capable and dedicated scientist: she became the first to study the full sweep of Siachen, which meant cataloguing the glacier's biological and geological diversity, naming its unreckoned peaks, and measuring its contours—work that revealed it as the world's longest known glacier beyond the polar regions. And yet she gained greater renown for suffragette stunts on mountaintops than these scholarly contributions—less a reflection of the quality of her science, perhaps, than of the fact of her being female in an era when explorers weren't. Whatever the case, her ideas were overlooked in contemporary geographic literature, her mapping criticized as inaccurate, and her surveying nomenclature almost entirely discarded. Even modern historians have uncharitably dismissed Fanny as "farcical," "amateurish," and "responsible for introducing a slight note of comedy into the awe-inspiring world of the high peaks."

As I read these criticisms in the Bodleian Library, or the "Bod," as students called it, I thought back to sneaking into Tibet, where a chubby Chinese policeman had handed me not a stiff fine, or even stiffer handcuffs, but cucumbers. Cucumbers! Only people who'd never actually travelled to the Himalaya could claim humour has no place there. Compared to the preening self-importance of most early Himalayan explorers, Fanny brought a refreshing dose of flair and whimsy to the highest altitudes. I admired her unlimited verve and refusal to be demure. Calling Siachen "the Rose" was, perhaps, a bit much, but a glacier is still a glacier by any other name. Just as the Tibetan Plateau, whether you call it Bod or Xizang, heaven or hell, is still a sky-raking tumult of rock and ice and turquoise water—the kind of landscape that, as even Fanny confessed, "was ever tightening its grip on my soul."

The plateau felt even closer when His Holiness the Dalai Lama came to town. I weaved through crowds to find my seat in the Sheldonian Theatre, where hundreds of students and professors leaned forward to glimpse the smiling monk in the middle of the room. Here was someone worshipped as a god since childhood, someone forced to flee a homeland so transformed by the Chinese that if he ever made it back to Lhasa (unlikely, given the Chinese government deems him a terrorist), he probably wouldn't recognize the Potala Palace, where a four-lane paved road has replaced the front lawn where pilgrims used to gather. And yet the Dalai Lama sat in the theatre beaming through thick-rimmed glasses, giggling at his own jokes. The frivolity of his laughter against the hard facts of his life made him seem a living koan, a riddle in the Zen Buddhist tradition that demonstrates the inadequacy of logic and provokes enlightenment—or, in my case, giggling in turn. I didn't think the Dalai Lama would mind.

He introduced himself as "just a simple monk" and gave a short talk on kindness. Afterwards a bunch of scholars, in typical Oxford fashion, asked questions that weren't really questions but statements designed to reveal the asker's own erudition. Jamie had sometimes fallen into this, for he could argue any issue from any angle so persuasively that you couldn't tell how he felt about anything. Perhaps he himself didn't know. "You don't believe in the basic rights of people to food, water, education, jobs?" my friend had grilled him over dinner one night. I knew his dismissal of development wasn't as glib as it sounded, that in questioning "progress"—our collective striving to make life easier and more comfortable for everyone—he was passionately interested in the relationship between suffering and the sublime. He hoped to get at the heart of why a drop of water in a desert tastes so sweet. But instead of opening up about any of this, Jamie had dodged my friend's questions with detached, scholarly feints of logic, which was easier for him. Frustratingly easy, from my perspective, but then I'd never figured out how to be anything but earnest.

The Dalai Lama had little patience for semantics. At one point a philosophy professor stood up and held forth on the distinction between "compassion" and "kindness." Noting that His Holiness had used the latter term throughout his talk, this professor praised the Dalai Lama for such a clever verbal strategy, given kindness was a more accessible concept for the masses than compassion, which had connotations of divinity, of excessive and unattainable virtuousness, and therefore seemed less authentic.

"Oh no!" The Dalai Lama giggled. "It is my English that is not authentic! Kindness, compassion, they are same. No strategy, ha ha! These are simple things, hmmm? Simple to say, harder to live."

I left the Sheldonian craving a thousand years to think over the Dalai Lama's words. Instead I settled for a long run through Port Meadow, a thousand-year-old commons where Buddha-bellied cows and sheep graze still. The tragedy that so famously afflicts the commons can be averted through mutual respect and restraint, or the sort of kindness the Dalai Lama was talking about: a basic empathy for others, the recognition that your desires matter no more and no less than anyone else's. By contrast, greed and ego— both on the individual and national scale—were the driving forces of exploration, with everyone gunning to claim all they could of the world before somebody beat them to it. Every week at Oxford I seemed to banish yet another exploratory idol from my pantheon, most recently Richard Hakluyt, the namesake for the Mars Society letter-writing prize I won as a teenager. Though Hakluyt wasn't technically an explorer himself, he was a loud evangelist for the European colonization of the "New World," which didn't turn out so well for the people already living there, a legacy now being repeated by the Chinese in Tibet. But almost as disturbing as such overt exploitation was the kind of exploration that had been initiated in total innocence and integrity—and led to disaster nonetheless.

A half-century after Fanny's expedition, for example, Siachen lost the honour of being the world's longest glacier (one in Tajikistan proved even longer) but it eventually gained the more dubious distinction of being the world's highest-altitude battlefield. After the Line of Control was drawn through contested Kashmir in 1972, dividing the territory between the newly designated nations of India and Pakistan, the boundary was terminated at survey point NJ9842, in the foothills south of Siachen, and from there vaguely extrapolated "thence north to the glaciers." Neither country much cared about Siachen, a place dismissed as a wasteland, and therefore exiled beyond the bounds of territorial ambition.

This situation began to change in the late 1970s and early 1980s, as mountaineers from Japan, England, and America requested permission from Pakistan—not India—to attempt peaks on the glacier, simply because it was easier to access Siachen from that country. The climbers didn't mean to choose sides, but even so their passport stamps implied Pakistan controlled all that ice.

What further riled India were foreign maps hinting that Siachen belonged to Pakistan. The original cartographic error can be traced to the United States Department of Defense's 1967 Tactical Pilotage Charts for Kashmir, which showed a dotted Line of Control angling "thence north to the glaciers" in such a way that Siachen was suggestively contained within Pakistan. Instead of jagging and twisting along the natural if squiggly ridges of the mountains, like the rest of Kashmir's Line of Control, the DOD line ran straight as a gunshot from NJ9842 to Karakoram Pass, once part of the ancient Silk Road. I've often wondered who made the fatal decision to draw that dotted line, and whether the decision was made innocently, for purely navigational purposes (perhaps the Karakoram Pass made an ideal piloting landmark?) or out of a more surreptitious fidelity to Pakistan, a country long allied with the US military. In any case, the DOD lines were not official boundaries, yet other maps reproduced them without this crucial caveat, including the reputable world atlases of Rand McNally, the Oxford Encyclopedia, and National Geographic. When a colonel in the Indian army happened to meet some Germans who planned to raft the Indus River, he was shocked to see that their American-made map showed Siachen as effectively belonging to Pakistan. India invaded the glacier in 1984 to prevent those paper-based borders from becoming a reality, Pakistan responded by sending its own troops to Siachen, and so began an escalating altitude race to the staggering heights of human absurdity.

Ever since, soldiers from both armies have lived year-round at elevations where few mountaineers dare linger. A ceasefire has been in effect since the early 1990s, but most casualties in the Siachen conflict result from natural hazards like avalanches and altitude sickness rather than enemy fire, meaning the death toll hasn't diminished much. Millions of dollars are spent daily to maintain troops on the glacier, and because it's too expensive to fly trash down the mountain, human waste and other refuse gets dumped in crevasses. After three decades of military occupation, the Siachen Glacier, a place one early explorer raved about as "indescribably grand, its unrelieved and elemental savagery producing in an unusual degree a feeling of exaltation and intense remoteness from humanity," has been reduced to what Indian army officials call "the world's highest and biggest garbage dump."

Why so much costly fuss over a far-flung chunk of ice? Because what the glacier lacks in strategic worth it makes up for in symbolic value, and neither country wants to lose face by losing Siachen, even as the standoff destroys the commons both wish to claim. In this respect the conflict over the glacier seemed itself a koan, and as I ran back to Oxford, which looked from Port Meadow like an alpine cluster of spires and towers, I wondered what kind of enlightenment a war-torn wilderness could possibly provoke.

Maybe reading about the Himalaya brought out my usual migratory instincts. Maybe Jamie wasn't goofy enough, or I didn't bring out his lighter side just as he didn't bring out mine, and so our relationship existed purely on the deep, searching plane where I spent too much time as it was, and therefore felt to me unsustainably intense. "Maybe we love each other best in words," I mused to Kim as we set the table at Holywell Ford, a quaint graduate residence at Magdalen College that looked roughly how I pictured Darwin's countryside retreat, with vine-tangled stone walls and a bucolic forest setting. Fortunately this cottage was filled with far more music, poetry, and laughter thanks to the irreverent crew from South Africa, Australia, Mexico, America, and Canada who gathered there weekly to cook dinner, drink wine out of jars (the shared kitchen lacked glasses), and debate serious academic questions like "If the field of global public health was a celebrity, who would it be?"

What united us "Jar Kids" across countries and academic disciplines was our mutual appreciation for the absurd. Jamie operated on a different wavelength entirely, and it was probably telling that he didn't join us often at Holywell Ford, that I spent more and more time on my bike, or reading and writing in the Bod, or executing pranks with the Jar Kids, such as filching our friend's snow globe of the Virgin Fatima and taking "selfies" of the sparkly Catholic saint around campus, which Fatima posted to her very own Facebook profile. Between dinner parties at Holywell Ford I subsisted on cheap muesli from Tesco, free coffee and cake from Rhodes House, and expired Clif Bars that I bought in bulk off eBay to save money for discount flights to Morocco and Norway, and later India, Chile, and Nepal. Graduate seminars in the history of science at Oxford met once a week, leaving the rest of my days free for adventure as long as I brought along my books, which were just as easily studied in a tent as in a library. Life had never seemed so open and reeling with possibility—except when I was with Jamie, which was less his fault and more a function of my chronic restlessness. Why hang out with the same person week after week when the world is calling? In any case, not long after he suggested we move in together, I decided I wanted out entirely.

After we broke up I went hiking in Wales for a weekend, craving a landscape large enough for my despair—over the fact that I would never find another letter in my pidge from Jamie, among other reasons I'd genuinely miss him in my life. Was I making a huge mistake? But I was wrong, if only about the letter. When I returned to Oxford, on the brink of getting back together with Jamie, he said he'd written something for me, but I could only read it on two conditions: that I remember he was very upset when he wrote it, and that I give him a copy because it belonged in the archives of his life. Then he handed me not the usual thin envelope I was used to finding in my pidge, but a subpoena-like stack of white sheets.

I brought the sixteen-page, hand-scrawled missive to Kim's room because I needed to not be alone when I read it. It took me two hours. The letter was a dramatic exegesis of our relationship, how we were meant for each other, how I'd messed it all up with my heart of stone, "which even a lawyer couldn't crack." On that point he was right, but where logic had failed, some levity might've helped. I returned the letter to Jamie without making a copy. The original, for his archives.

The next day he called and said he wanted to talk. I didn't answer the phone or return his call, but I ran into him a few days later on Catte Street. "Wait," he pleaded. "I'll be right back." He returned with a poster of Mount Everest that he'd gotten signed by a renowned alpinist at a lecture in London, an event we were supposed to attend together. "A peace offering," he explained, his face paler than ever. Mine probably looked the same. I stared at the poster with a mountain-shaped hole in my heart and longed to be anywhere in that wildness, that slant of light so intense you could lean into it and be held.

Reading about the Himalaya was the next best thing. I decided to write my master's thesis on the Siachen Glacier, detailing the soap-opera saga of exploration and geopolitics that had played out on its ice, but also exploring the possibility of scientific collaboration as a potential solution to the conflict. If scientific expeditions and their subsequent maps had led the way into war on Siachen, however inadvertently, could science, that supposedly neutral, non-nationalistic pursuit of truth, also lead the way out?

A dear friend tipped me off to the idea of "scientific

peacekeeping" as a potential way to resolve the Siachen conflict, namely by rendering the icy battlefield into a demilitarized buffer zone dedicated exclusively to scientific research. It was a dreamy notion, but people had done it before: in Antarctica, for example, where countries with competing territorial claims formally agreed to disagree about who owned the continent and collectively set it aside for science. If such a treaty could happen for a cold, remote, uninhabited continent coveted by dozens of countries, why not for a cold, remote, uninhabited glacier few people outside of India and Pakistan had even heard of? Not that obscurity or isolation is an anti-dote to desire; sometimes those very qualities amplify longing, as the bleak yet coveted Aksai Chin attests. Even so, a science-focused treaty on Siachen seemed reasonable, a way to defuse a senseless military standoff and return the glacier to its original unowned state.

The history of exploration, as I'd learned at Oxford, was basically synonymous with imperial expansion and indigenous repression—a rather cringe-worthy legacy for an endeavour I once deemed so essentially wondrous and searching. Science and exploration as a force for peace in contested frontiers seemed to offer a kind of redemption. I was so compelled by the work—and the Himalaya—that I seriously entertained the idea of staying at Oxford for a doctorate. After all, the Rhodes offered me another year of funding, and the laboratory at MIT wasn't going anywhere. It wasn't that I wanted to be a historian of science, exactly; what I craved was the life of reading, wandering, and writing that studying history at Oxford made possible.

Knowing I needed a "distinction" result in my master's degree to qualify for the doctorate program, I worked harder on my thesis than I'd ever worked before. Strangely, it didn't feel draining the way laboratory science and problem sets always had. Reading and writing mysteriously gave me more energy than I put into them. I would ride my bike around Oxford for hours each morning, then work all day and late into the night on my thesis, furiously awake and full of questions.

"Well done, a fascinating piece of work," Corsi raved after reading an early draft in my second year at Oxford. We talked about Fanny, the Aksai Chin, the way maps confuse possession for control. We discussed the Outer Space Treaty, which was one of my case studies for Siachen, and the seductive delusion of a "final" frontier. I was just working up the courage to ask if he'd be willing to supervise my doctoral studies, in which I hoped to widen and deepen this work on contested borderlands, when Corsi's tone changed. "But Kate, I must say," he began in a manner ominously less operatic. "I worry your thesis may not qualify as the *history* of science. So much of it concerns the present, the future, no?"

Hesitation, it turns out, is the hardest frontier to cross. I was so shaken by Corsi's words that I lost all faith in my project, the idea of staying on at Oxford for a doctorate in the history of science. I debated finishing the Silk Road immediately, biking contested borderlands from Turkey back to Tibet and on to Siachen—a way of finishing what I'd started and ground-truthing what I'd studied. I didn't need academia as my excuse to read, wander, and write. But Mel was busy with her own master's degree in community development, and I was too good at school, in every doomed sense. After being on an achievement bender most of my life, the prospect of withdrawal, of doing anything without external approval, or better yet acclamation, kept me obediently between lines I couldn't even recognize as lines. Isn't that the final, most forceful triumph of borders? The way they make us accept as real and substantial what we can't actually see?

In any case, I went back to my room and applied to MIT.

That spring I submitted my seemingly hopeless thesis on Siachen. A month or so later I wrote my final exams, a surreal, uniquely

Oxonian experience in which students must wear a mortarboard or soft cap, a black academic gown known as a "sub fusc," and a flower pinned to the gown where a breast pocket would be. But not just any flower: you must wear a white carnation for your first examination and a pink one for all but your last, when you wear a red carnation, which signals to the mob waiting outside Exam Schools that you're ready to be stormed with champagne, whipped cream, and glitter. An oddly satisfying if dizzying end to a master's degree, itself an experience not unlike being shaken in a snow globe until the glass breaks and releases you into the bright air, drenched and sparkling, where you celebrate by drinking wine out of jars with your friends.

Shortly before I left Oxford, Jamie invited me over for dinner for old times' sake. Over Greek salad and wine the two of us talked our way back to the Silk Road, Egypt, a remote Arctic island long ago in a galaxy far away, or was it Stratford? In a strange congruence, we realized we'd both studied dumps for our master's theses: on the Siachen Glacier in my case, and in Cairo in his, among the Zabbaleen, which means "garbage people" in Arabic. Though we'd each written critiques against a certain kind of progress, or at least against the natural and spiritual wreckage that so often accompanies it, Jamie observed that we'd used opposite methods: he'd embedded himself in civilization's deepest, most degenerate core and projected into the future, while I'd withdrawn to civilization's outer, mostly untouched edge in the Himalaya and projected into the past (if not far enough, according to Corsi, but I kept this to myself). I stared at Jamie, stunned, for I suddenly guessed why our fascination with each other was mutual, sincere, and finally incommensurable: we looked at the world through the different ends of a telescope.

By tradition you don't enrol at Oxford, you "come up," and

graduating is called "going down." Not since biking onto and off the Tibetan Plateau had this terminology been so existentially apt. In a blur I packed my bags and my bike and caught a flight to Ontario. My brother Dave drove me to Massachusetts, where a middle school teacher named Sara welcomed me into the house I would share with her and two others, including a pale, furtive woman who rarely emerged from her room and then only mincingly, on tiptoe. As we passed her closed door Sara whispered, as if by way of explanation, "She also goes to MIT." Then she directed me to my room, a crawlspace that would just barely fit my futon mattress if I bent it like a hot dog bun. "Oh yeah," Sara said as she left me to unpack. "Some mail came for you."

What does it mean when you build your own walls? You have no one to blame but yourself for inhabiting them. My hands shook as I tore open the thin white envelope from Oxford, which I knew contained my degree results. I reminded myself that it didn't matter, that the history of science was just a detour, a side trip, a delirium of work that felt, confusingly, like deep play.

I read the slip of paper. Read it again. Several worlds slid past me at different speeds.

MIT started promisingly enough, despite the gutting regret brought on by the unexpected "distinction" on my master's degree from Oxford. I'd barely unpacked my bags when my Ph.D. adviser, Dr. Tanja Bosak, sent me with her other students to Yellowstone National Park for fieldwork. Born in Croatia and trained at Caltech, Tanja was a brilliant scholar, a kind and encouraging mentor, and the sort of scientist who mostly preferred studying the natural world from within the walls of a laboratory, a preference I realized belatedly. In what should've been a warning sign, she didn't seem especially heartbroken about missing a week of fieldwork in the

Wyoming backcountry. "Don't forget the bear spray," she reminded us, her eyes bright with the barely suppressed cheer of someone relieved at being left behind.

Every day the other students and I hiked around Yellowstone's hot springs in search of samples, collecting slimy blobs of microbial mats from boiling puddles. Every night we ate dinner around a bonfire beneath the stars, watching geysers spill into the sky like the source of the Milky Way. I could get used to this, I remember thinking as I tucked into my tent each night. But the next time I went camping in the name of science was in my office at MIT.

When the alarm blared at 6:00 a.m. I sat up in my sleeping bag and bashed my forehead against the desk above me. I'd stayed late working on a problem set that, among other things, asked me to calculate the flux of marine snow to the ocean floor. "What's marine snow?" I'd furtively questioned a fellow student when the professor handed out the assignment. I pictured polar blizzards, Shackleton's ship crushed in the Southern Ocean, Nansen drifting in the Northwest Passage at the pace of pack ice.

"Planktonic fecal pellets," my classmate clarified.

By now the bait-and-switch was nearly complete. After returning from Wyoming, Tanja had called me into her office to explain that she had enough students working on the Yellowstone samples. Rather than studying life in extreme environments, how about I focus my Ph.D. on laboratory studies of molecular biomarkers instead? When microbes die, she explained, pieces of them survive as fossils under certain geologic conditions. Fat in particular is as difficult to get rid of in death as in life, and microbial lipids, such as polycyclic triterpenoids, can stick around in sedimentary rocks for billions of years. By comparing these molecular fossils to the lipids of modern micro-organisms, such as *Rhodospirillum rubrum*, we can piece together the history and evolution of life on Earth. Billions of years ago our home planet was by all measures an alien world, with different contortions of continents and a dearth of oxygen. "Using similar techniques," Tanja added, "it might be possible to search for traces of life on Mars."

She won me over, if reluctantly, at the mention of the red planet, and I moved into the lab for the long haul.

Why Mars still exerted such an obsessive pull on me I can't really explain. I was caught in its gravity, I suppose, though any MIT physics student would dismiss that possibility, given the red planet has less mass than the Earth, making its pull a third weaker. Then again, literature grapples with forces science can't detect or express. "We tell ourselves stories in order to live," wrote Joan Didion, and years later, reading those words, I recognized that voyaging to Mars was precisely that for me: a survival narrative I'd cleaved to in a world that seemed mapped and tamed. A fiction awaiting its paradigm shift—to another fiction, perhaps, but a better or at least wilder one.

But even more than a story, if I'm being honest, Mars was almost a god to me, the galvanizing force of my life, and in that sense MIT was a firm step closer to the divine: after all, the university has graduated more astronauts than any other in the world. The problem was that I wasn't sure anymore if I wanted to go. Especially if the statistically remote possibility of being selected as a Martian colonist required, in the intervening decades, that I wear clammy purple nitrile gloves and squint down a modified spyglass—not at the moons of Jupiter or even the King's Arms pub, but at mindless battalions of microbes in a petri dish.

Which isn't to say wildness and mystery can't be found on all scales, from bacteria to black holes. The poet Blake saw the universe in a grain of sand. A vantage tinier than my pinky fingertip yielded to Galileo an infinite vista, pinpricks of light that shattered the godlike perfection and glasslike immutability of the heavens. And so I talked myself into staying in the lab even as I suffered the elder Darwin's data-driven loss of happiness, and the words of the younger Darwin rang in my ears, urging me "to take all chances and to start on travels by land if possible, if otherwise on a long voyage." Instead I spent my days parsing apart the microbial equivalent of cholesterol, and my nights solving problem sets on the flux of excrement to the ocean floor. Was it any wonder I began to question the meaning of life?

"Kate, Kate, Kate," Tanja would cluck in her lilting Croatian accent. "That's not a scientific question." She encouraged me to pose more experimentally tractable queries, such as what kinds of polycyclic triterpenoids are produced by *Rhodospirillum rubrum* when incubated with varying concentrations of sulphur and oxygen?

I had to admit I didn't know.

"Well, then," she would say, beaming, "there's a whole laboratory behind that door in which you can seek the answer!"

A year came and went in a numbing blur of problem sets and experiments. Other than a two-week summer holiday in India, during which I tried and failed to see the Siachen Glacier, my sleeping bag lived in my office and saw a lot of use. I adored Tanja and didn't want to let her down, so I tried not to think too much about Oxford, about detours versus destinations. I dated around out of the most flailing loneliness I'd ever felt, hoping the right relationship could remedy the wrong life. I even second-guessed breaking up with Jamie, though in England, where I was happy, I'd rarely doubted the decision. "One word of praise for my writing from you," he'd written to me, "means more to me than my Oxford degree." But he hadn't failed out after all and was completing his doctorate in Development Studies. Most of the Jar Kids had stayed on at Oxford to do the same. The fairy tale continued without me.

Life only made sense now when I rode my bicycle. I'd taken up cyclocross and mountain bike racing at MIT with my usual reckless avidity, which worried Tanja because the university's cycling team members often took twice as long to finish their Ph.D.'s. In truth, I suspect racing was the main thing keeping us going: a regular dose of endorphins that made the daily pressure and tedium of lab experiments and problem sets bearable. In the total focus required to stay upright on the rutted, twisting trails of the race courses, I could almost forget the fact of grad school. I went on to medal in the national collegiate mountain biking and cyclocross championships, but competition for me was less about beating others than defying the tyranny of my ruling, rational self: when my mind screamed *stop*, some more ancient and cardiac instinct urged *go*.

So I went, around and around, pretending the race course was a tightly spooled Silk Road. How many laps back to Lhasa? How many pedal strokes to where Siachen noses coldly into the Nubra Valley? It was a relief to give anything my all again, to strain toward something as neat and tangible as a finish line. Was that Fanny Bullock Workman in tweed on my left? And Alexandra David-Néel just ahead, prayer beads rattling on her bike frame? I accelerated to catch them but they always disappeared after a few laps, though I couldn't see any turnoffs, any alternate paths or escape routes, and the faceless crowd roared. Eventually the world would go mute and I'd be alone again, the elder Darwin mechanically pacing his cottage, or an astronaut looping endlessly in low Earth orbit, digging my own tracks a little deeper each time around.

What finally sent me spinning back to the Silk Road was a meeting with Dr. Maria Zuber, then the chair of the Earth, Atmospheric and Planetary Sciences department at MIT. Slim, resolute, and formidably smart, Zuber always looked to me as if she were leaning into a brisk wind. A geophysicist by training, she'd led or been involved with several NASA missions to map various asteroids and planets, including the red one. This was my motivation for meeting with her, to see about doing a different research project on Mars. That planet had once made me want to be a scientist, after all, and I desperately hoped it would do so again.

"So," she began sharply as I settled into her office in Building 54. Most buildings at MIT are known by numbers, not names, a quirk that struck me as charming when I first arrived and chilling when I left. "What do you plan to do with your life?"

"I've always wanted to be an explorer," I confessed, and instantly regretted it. Such an aspiration sounds whimsical from a seventeen-year-old high schooler with a dream, but worrying from a twenty-seven-year-old Ph.D. student with a graduation deadline.

To my surprise, Zuber responded with enthusiasm. "Wonderful! You've picked a great time to be alive," she exulted, as if I'd had a choice in the matter. "It's the Age of Discovery, whatever the history books say." She tidied a minor avalanche of papers on her desk. "I mean, how amazing is it that we can sit at a desk, right here in an office, and explore Mars from a computer screen?"

I winced and hoped it didn't show.

"Just imagine," she continued devastatingly. "Magellan had to sail in rough seas for months, even years on end. Risking scurvy, cannibals, strange diseases, who knows! But today we can wander another world with our feet up on a desk and a Diet Coke in hand. There's never been a better time to be an explorer."

On her desk was indeed a soda can, slightly dented in the middle, as though it had been clenched tight in a spasm of late-night frustration. Zuber's feet were firmly planted on the floor and it was hard to imagine them anywhere but. I thought about the NASA space probe named after the Portuguese explorer she'd mentioned: Magellan was launched in 1989 to map the surface of Venus using radar to penetrate the planet's obscuring clouds, allowing scientists—among them Zuber—to study its volcanism and tectonics. After five years in orbit, roughly the time it takes to finish a doctorate, NASA deliberately incinerated the probe in Venus's dense atmosphere, a series of events that suddenly struck me as a parable.

I forced my attention back to Zuber, who was discussing other matters now, plans and logistics for the next semester, research methodologies. I listened and nodded, eyeing a framed map of Mars on the wall. It was slightly crooked and covered in names. Near it a few plants wilted on a windowsill.

". . . and then we munch cucumbers," Zuber finished.

"Excuse me?" I said, startled.

"And then we crunch the numbers," she repeated. "Simple as that."

I politely thanked Zuber for meeting with me and walked out of her office. And then I left the laboratory and launched on a long voyage.

PART TWO

〜

Never to get lost is not to live, not to know how to get lost
brings you to destruction, and somewhere in the
terra incognita in between lies a life of discovery.

REBECCA SOLNIT,
A FIELD GUIDE TO GETTING LOST

4.

UNDERCURRENTS

Black Sea

T he Silk Road is old news now, fable uncombed from fact. Once a dynamic flux of trade and ideas between Europe and Asia, this bygone caravan route now mostly traffics in drugs and violence at worst, myths and souvenirs at best—wares Mel and I hoped to avoid as we set off down it on bicycles. Or more accurately, *with* bicycles. I tried to muster an air of dignity despite wearing a helmet while pushing, and not pedalling, my overburdened wheels through the crowded streets of Istanbul. "Don't crash!" a lanky Turkish teenager mocked in English. I pretended not to understand.

The sky was dull and shapeless that January morning. Camera-slung tourists strolled past pyramids of turmeric and paprika in the Ottoman-era Spice Bazaar. Sizzling torsos of marbled meat pirouetted in restaurant windows. Outside the New Mosque, a man with a drooping white moustache flogged Turkish flags by draping the red fabric cape-like around him, perhaps hinting at nationalism's superpowers or offering himself as part of a package deal. Noticing my stare, the man shouted something at me, startling a flock of pigeons into flight. Fat grey birds scattered in the square like a toss of ball bearings.

I took off my helmet and slung it from my handlebars. One thing certainly hadn't changed in the epochs since Istanbul was Constantinople, and before that Byzantium: the place was a bustling shopping mall. Mel and I wheeled our bikes through the crowds toward the Bosporus, a sinuous strait cleaving Istanbul across two continents as it flows for twenty miles from the Black Sea into the Sea of Marmara and eventually the Mediterranean. We wandered past fishermen casting long lines into the soupy green channel, past bait stands selling bouquets of startled-looking silver anchovies in plastic cups. We found the ferry ticket booth, waited in line, and walked our bikes over a gangway as the call to prayer warbled across the water.

Mel and I leaned on the railing of the ferry and watched Istanbul fade behind us, with fog swallowing it dome by spire by billboard. I wasn't sorry to see it go. If the Silk Road's legendary trading hubs are now mostly reduced to rubble, or modernized beyond the reach of nostalgia, the hinterlands that Marco Polo cursed still exist much as they did millennia ago: deserts that slow all travel to a camel's slog, mountains that ramp into solitudes of ice and sky. Or so I hoped. For now I gave Istanbul a last glance. If all went well, we wouldn't see this many people in one place for almost a year.

Clouds pinched the sky. The air smelled of brine and coal smoke. Mel ducked inside to warm up just as a Turkish business-man came on deck and stood a polite distance away, puffing on a cigarette. He was fifty-something, stout, and stubbled, with a face that seemed on the verge of a yawn. "Forget biking the Black Sea," he told me, staring into the smudged mirror of the Bosporus, as if long ago he'd lost something in that murky water and expected it to resurface any second. "It's winter, very many rain," he said. "You must go south. Cappadocia, Konya, the Aegean . . ."

I smiled at the man and shrugged. We'd been warned the same thing by others, but how bad could winter biking at sea level be? Give me storms and scurvy any day over a slow, pale death by computer screen and Diet Coke. "I'm breaking up with Mars," I'd declared to Mel on the phone shortly after my meeting with Zuber. "The long distance thing just isn't working for me."

It took a while to end it, as these things do, but eventually I was free and so was Mel, and we decided to finish the Silk Road we'd started five years earlier. Which meant biking over roughly a year from Turkey back to Tibet and on to the Siachen Glacier—a place Polo didn't actually visit but surely would've despised for its vast-ness, severity, and glaring lack of marketable commodities. Beyond avenging my childhood ideals of explorers, and figuring out how to be one myself, I wanted to bike the Silk Road as a practical extension of my thesis at Oxford: to study how borders make and break what is wild in the world, from mountain ranges to people's minds, and how science, or more specifically wilderness conserva-tion, might bridge those divides. So there I was, rich in unemploy-able university degrees, poor in cash, with few possessions to my name besides a tent, a bicycle, and some books. I felt great about my life decisions, until I felt terrified.

The propellers churned a stripe of turquoise into the smoky, emerald waters of the strait. Europe was to my left, Asia to my right, and below me a fluid borderland whose depths I couldn't fathom. The name Bosporus is Greek for "ox ford"—but was I coming up or going down? Too early to say, or perhaps too simple a distinction.

In 1680 a young Italian nobleman named Luigi Ferdinando Marsigli tossed a weighted line into the Bosporus and realized, as it arced first away then toward him, what Turkish fisherman already knew: the strait flows two ways at once. The history of science and exploration is full of wealthy foreigners winning fame and glory for "discovering" the sorts of things locals already knew, but Marsigli deserves credit not for realizing the Bosporus has an undercurrent, but for figuring out why: a salinity difference between Black and Mediterranean sea water. As rivers pour freshwater into the Black Sea and eventually out the Bosporus Strait, denser, saltier water from the Mediterranean flows in to fill that space. The ferry was muscling against the surface current of the Bosporus on its way to the Black Sea, but would effortlessly surge there if it could only dive a few dozen metres deeper.

The businessman stubbed out his cigarette and went below deck. When I went inside a few minutes later I found Mel chatting with a genial-looking young man with round cheeks. After a double-take, I realized it was Jeremy, a classmate from elementary school whom I hadn't seen since the ninth grade. He and his fiancée, Kerri, were enjoying a holiday in Istanbul and happened to board the same ferry as us, which meant we had a hometown witness when we missed our stop.

"So where are you two getting off?" Jeremy asked. When Mel told him "Beynou," he looked concerned. "Uh, didn't we just pass that place?"

We sheepishly walked our bikes down the plank to Anadolu

Kavagi, the final stop on the ferry, a small village barnacled on the ragged bank of Asia Minor. In a dockside restaurant we shared a greasy seafood lunch with Jeremy and Kerri, then lingered awkwardly, hoping the two of them would leave: we didn't want to reveal that our hastily printed-out Google maps only showed the route from Beynou. Although we also carried a foldable map of the Silk Road, it was so large-scale as to be useless for actual navigation. Instead of saying goodbye, though, Jeremy kindly offered to film us setting off on our journey. Unable to muster further reasons for delay, we mounted our bikes and started pedalling to Siachen.

A few seconds later we split ways. Mel veered right while I turned left at the first intersection, a parting beautifully timed and choreographed but for the minor detail that it wasn't planned. We hadn't conferred on which way to turn, though it was anyone's guess given we were already off our map. A fitting start to the ride, I suppose, for the Silk Road is as knotted and intricate as the Turkish carpets I'd ogled in Istanbul's Grand Bazaar before deciding it was too early in the trip to buy souvenirs. Instead I bought baklava.

"Just a warm-up lap!" I hollered to Jeremy as I circled back to meet Mel, trying to look nonchalant and fit despite the betraying plumes of mist exiting my mouth. The only way to prepare for biking the Silk Road every day, all day, for nearly a year—or so I'd lazily reasoned before the trip—was to bike every day, all day, for nearly a year. Given this hadn't exactly been feasible, my training regimen had mostly consisted of last-minute bulking up on Turkish sweets.

I only remember fragments from that longed-for return to the Silk Road. Scenes like jump-cuts in a film, edited in a way that conceals the connecting story, perhaps because we really did record that first day and many that followed with a camcorder, hoping to make a documentary film about the expedition. Straight out of town we wheezed up a cobbled road, set up the tripod, hit the record button, biked down again, and wheezed back up all over again to capture us doing so on camera. The rest is a bit of a blur. At one point I briefly napped on a couch mouldering on a sidewalk, and at another the laptop Mel carried for video editing purposes flew off her rear rack on a descent and somehow outpaced the bike, so that she nearly ran over it. The leafless trees and bare fields lining the road looked pale and wrinkled, and the earth gave off the fresh aroma that follows rain. As dusk approached, we stopped at a roadside restaurant because we hadn't yet found a gas station where we could fill our cookstove fuel bottles, meaning the ludicrous quantities of noodles we were lugging were for the moment useless ballast.

The restaurant was closed, but the owner invited us to camp on his lawn and offered us tea—or *çay*—prepared from a double-decker kettle that we soon learned was universal on wood stoves across Turkey. The bottom kettle contained boiling water and the top had tea steeped to the point of sludge. He poured an inch of the latter into two tiny, slender glasses, then diluted this with hot water from the lower kettle, which he poured in a long, perfectly aimed rope. With a final ceremonial flourish, he plunked two sugar cubes into each glass. As we sipped this supersaturated liquid on an empty stomach, the owner engaged us in a deep, soulful conversation, or so I guessed from his earnest demeanour and emphatic hand gestures. I stared at him dumbly while Mel nodded and murmured "ahh" at sympathetic intervals. As we walked away, I asked her what he'd said. "No clue!" she answered.

Some dogs nearby barked at a passing car, then one shuffled over to our shiny new tunnel tent and peed on the door before we could stop him. We gingerly zipped open the Glow-worm, as we'd nicknamed our bright red abode, and crawled through the vestibule to the main compartment. "Please remove your high heels," Mel intoned with an Italian accent as she pulled off her sweaty boots, mimicking the Air Italia in-flight announcement on our connecting flight from Rome to Istanbul. The airline was clearly accustomed to a different, less adventurous clientele, judging from that announcement, as well as the fact that they misplaced Mel's bike. We'd felt lost in transit ourselves as we waited nearly a week for it to show up in Istanbul, our days dizzy with errands, emails, and shopping excursions to procure last-minute supplies with our fast-diminishing pool of money. By pitching this expedition to sponsors and granting agencies as a journey to explore wilderness conservation across borders, we'd managed to rally some funding and gear for the Silk Road, but even so we were thousands of dollars short of what we needed to get to India.

Not that it mattered now, in the sense that Mel and I had committed to this road and couldn't change anything. But I had trouble sleeping that first night on the Silk Road. Or maybe the sugar and caffeine somersaulting in my veins was to blame. Either way, math obsessed me in a way it never had at MIT. I did some quick mental calculations as headlights from a passing car glanced over the tent, making the interior glow as though it were sunrise or sunset. We'd biked for three hours and covered a measly six miles, in part because we'd doubled back so often for filming. At this pace, I realized with a sinking feeling, we'd reach Siachen in approximately three years.

Mel and I had set off to finish the Silk Road from Istanbul because we figured winter in Asia Minor would be less daunting than in the Himalaya. But the Turkish businessman was right. Freezing rain on the Black Sea coast wasn't just passing weather; it was a regime.

For the next month our whole world conspired to water. The

sky was fathoms deep with it, abyssal as the sea below. The sun slouched low on the horizon while I hunched on my bike in much the same way. According to Strabo, an Anatolian geographer born in 63 BCE, the Black Sea was formerly known as Axenos, meaning "inhospitable," named for its fierce tribes and wintry storms. The sea represented the fringe of the known world to the ancient Greeks, beyond which lay the realm of fire-breathing bulls, guardian serpents, and dragons whose teeth, when planted like seeds, were rumoured to grow fully-formed giants. Later the Black Sea was renamed Euxinos, or "friendly to strangers," a baptism as quaint and absurd as Fanny dubbing Siachen "the Rose." But if weather on the Black Sea wasn't hospitable, the people who lived there were, and night after night Mel and I found ourselves invited to stay with families.

Inside one home near Sinop, the wood stove threw heat like a small sun. In an armchair near it a little boy cuddled on his grandpa's lap, the tiny ship of his body harboured in a cove of wool and wisdom. His grandma lumbered over to the couch where Mel and I sat, and grinned with her whole face but only two teeth, matching ingots of spit-shined gold. On the wall was a portrait of Ataturk, "father of the Turkish nation," who in the wake of the Ottoman Empire established the separation of religion and state in Turkey, and now adorned household walls like a god himself.

After a hearty dinner of lentil soup, bread, and salad, Mel and I yawned and rubbed our eyes, hinting at our tiredness, hoping we'd all go to bed soon. Instead, we squashed into cars and drove into town. It was cousin Hande's tenth birthday party, a teenage boy explained, and we were going to her family's house to celebrate. Upon arrival the men dispersed to the local tea salon, leaving the women and children to laugh, talk, and eat in a very cramped living room. Soon the air was smoggy with heat and our trapped exhalations; the thermostat read twenty-nine degrees Celsius. Mel and I sat dripping in our thermal underwear and fleece pants, clothing more suited to weathering a blizzard than a Turkish birthday party.

Second dinner was served. Though first dinner had delivered more calories than a week's worth of instant noodles, I tucked into a plate of baklava like it was my job. The party swirled incomprehensibly around us, a chaos of balloons and Turks and jokes we didn't get but giggled at anyway. Techno music videos blared from a television, furnishing the party with a clubbish atmosphere even the gold-mouthed grandmother seemed to enjoy, for she nodded her shawled head to its hip, rhythmic pulse. The temperature kept rising. Someone on a Skype call aimed the video camera in our direction, so Mel and I gamely waved at the smiling, pixelated strangers on the screen and saw ourselves mirrored back with a slight but telling delay: two sweaty-faced foreigners lost in translation at a Turkish party.

Hande took advantage of my distraction on Skype by painting my fingertips with hot pink nail polish. It wasn't my style, but who could say no to a birthday girl? More accurately, who could say no in Turkish? I didn't dare. People here rarely said no at all, deeming it too blunt and dismissive a term. Instead, they said *yok*, meaning none, not existing, not here. A word as thick and satisfying on the tongue as baklava, though its implications for us were far less sweet. How far to India? *Yok.* Would the rain ever stop? *Yok.* In the name of all that is good in the world, where was the sun? Most emphatically *yok.*

The next morning I woke up with a baklava hangover and ate some more at breakfast in hopes of a cure. We dawdled in the house, reluctant to get back on the bikes. I sat near the wood stove, jotting notes in my journal, while Mel sat with the little girl and worked through grammar exercises in a Turkish-English textbook. The prompts were so bizarre that I was sure Mel was making them up. "You don't know Michael Jackson. Do you?" she read. "No," the little girl answered solemnly.

My friend finally sighed in a way that signalled it was time to go. As we slowly packed up, our host father protested that it was too cold and rainy, the road ahead far too dangerous. "Not just a road, a highway!" he warned, shaking his head. He said he would never let his daughter do such a trip, never.

Mel, all charm with her red curls, reasoned that if she wasn't travelling by bike across Turkey, she wouldn't have the chance to meet lovely people like him and his family.

At this he melted. "You are right, okay," he agreed warmly. "Yes, this is true!"

I silently cursed her for convincing him so easily. At the moment I wanted nothing more from life than to revel in the warmth and dryness of this cozy home, read books next to the fireplace, let my hot pink nails fade to a more modest tint, and never go back out there again.

What I craved was wilderness, and the Black Sea wasn't it. Every day people asked us what we thought of Turkey. "Your country is *chok güzel*, very beautiful," we told them, adding that we'd love to come back someday—in the summertime. But even warmth and sunshine probably couldn't redeem stretches of the Black Sea, particularly once the steep and winding country roads we'd started on morphed, as our host father had warned, into a heavily trafficked highway. The road plundered the seashore of any charm or elegance it once claimed. Most days we felt like we were biking through the scum on the rim of a giant bathtub.

The analogy is apt. Edged by six countries, and fed by rivers from twenty more, the Black Sea drains nearly a third of continental Europe. Since the sea's only outlet is the strangulated upper current of the Bosporus, as Marsigli showed, its deepest layers lie relatively still and stagnant. These bottom waters are poor in oxygen but rich in hydrogen sulphide, a colourless, poisonous gas that reeks of rotten eggs. Except for a few hardy microbial reefs, which subsist on methane seeps on the sea floor, little survives down there.

The road that edged it was similarly bleak. We pedalled a four-lane highway where nothing lived but speed and grief. We passed ditches floating with soda cans and dead dogs, torsos inflated like furred balloons. We passed a woman in a bus shelter with a face like a perpetual wince. We rode by a freshly splattered cat, then watched a second cat slink through the traffic to check on his mashed companion. We narrowly skirted a pile of dead anchovies on the road shoulder, and the smell of rotting fish stayed with us for miles. I thought about how caviar was once so plentiful, in fourteenth-century Byzantium, that it was considered the food of the poor. Centuries before that, Strabo reported that you could pull bonito, a small relative of tuna, out of the Bosporus with bare hands. Now I was more likely to pull out a plastic bottle.

Although the Black Sea's oxygenated shallows and undersea shelves once boiled with life, coastal cities in the nations surrounding it have been dumping pesticides, fertilizers, detergents, and poorly treated sewage into the shared borderland of its waters. This flush of nitrogen and phosphorus has triggered massive blooms of phytoplankton, which grow in vast, rippling sheets of crimson goo that shield sea water from sunlight. When the blooms die off and decompose, huge quantities of oxygen are consumed, leaving surface waters nearly as anoxic and sterile as the Black Sea's depths. Only a few invasive species thrive in such conditions, among them the rapa whelk, a Japanese snail that has decimated bivalve diversity in the Black Sea. Beaches are strewn with punctured seashells, the tiny holes marking where small whelks drilled into the carapace, injected digestive enzymes, and slurped out the liquefied flesh. Large whelks don't waste their time drilling but instead pry bivalves open with their one creepy muscled limb.

It was enough to make MIT seem appealing. If I was doomed to spend my life with anoxic microbes and alien species, at least laboratories lacked rain and roadkill. Turkey was nothing like I'd imagined, nothing like the Silk Road I'd dreamed of and talked about finishing for so long. If someone had offered me a spacesuit at that point in the bike ride, I would have accepted it gratefully, relieved for any kind of protective barrier between me and weather and traffic on the Black Sea. At what point was I running away from life, and at what point was I running toward it? The distinction suddenly struck me as crucial and troubling.

A truck veered so close to Mel that its draft sucked her off the road shoulder and onto the highway. Fortunately no other traffic was nearby. She veered back onto the shoulder, giving the truck driver the finger in the process, but he didn't see it or didn't care. The two of us differed only slightly in scale from the bugs flecking his truck's giant windshield. When I'd realized, as a kid, that explorers should put themselves at stake, I hadn't exactly had Turkish transport trucks in mind.

"And we chose this," I despaired. "We have no one to blame but ourselves."

"And Marco Polo," Mel added.

She had a point, though some historians doubt the Venetian merchant ever travelled beyond the Black Sea. Polo's name isn't mentioned in surviving Mongol or Chinese records, which seems odd for a high-level diplomat in the court of Kublai Khan.

He confused major Asian battles that happened years apart. He failed to mention the Great Wall, chopsticks, and other striking peculiarities of the region he supposedly called home for more than a decade. Because of these errors and omissions, some scholars, notably British historian Frances Wood, argue that Polo likely stopped travelling thousands of miles short of the Orient and that his stories were merely hearsay from fellow traders.

She might as well have accused Neil Armstrong of not landing on the moon. "I am condemned to a lifetime of Marco Polo," Wood noted ruefully in a lecture she gave a year after publishing her playfully subversive book, *Did Marco Polo Go to China?* "Before embarking upon what I had thought was an amusing little exercise in the cutting down (though not necessarily total demolition) of a legend, I had no idea how seriously Marco Polo is viewed." He is seriously viewed in the public imagination, at least, where the Venetian merchant has long been apotheosized as a household name—and one romantically synonymous with "explorer," though all Polo did was travel to lands new to him but old to others and write about what he saw. Could it be so simple? The idea gave me strange hope.

Some academics take Polo at his word, arguing that for whatever picky facts he got wrong—such as reporting twenty-four rather than thirteen arches on what is now called "Marco Polo Bridge" near modern Beijing—he got many other cultural and geographic details right, far more than can be ascribed to fluke or gossip. Plus Polo didn't even write his book; he dictated stories to Rustichello da Pisa while both men were imprisoned during the Venetian-Genoese Wars, and Rustichello compiled them into *The Description of the World*. Any errors or omissions in the manuscript, then, can be blamed on a bad ghostwriter.

I align myself with the believers, though for a rather

different reason: the book is frankly too boring to have been made up. If Marco Polo was such a fabulist, why does his magnum opus read like a guidebook written by a merchant for other merchants? His account of the Silk Road is so utilitarian, so oblivious to wonder and beauty, so obsessed with the bottom line. If Polo had written a dream-like sequence in the style of Italo Calvino's *Invisible Cities*, describing a Silk Road of possible futures and unforgettable pasts—territories of memory and desire, ruin and renewal—I might be less likely to believe the book. I would also love it more.

Of course, I'm judging Polo's work by modern literary standards. In his day and age, *The Description of the World* seemed as sensational as a science fiction novel, at least to its mostly European readers, who had never heard of cities with twelve thousand bridges, winds so hot they suffocated people and turned them to dust, or black stones and a black liquid that burned. His book would go on to be the most famous and influential travelogue of all time, spurring the likes of Columbus to seek shortcuts to Asia's treasure trove of gold and spices. When it was first published, though, the book mostly earned Polo the mocking nickname "Il Milione," or Marco of the Millions, reflecting the extravagant scale of his claims about the wealth and territories of Kublai Khan. People doubted the Venetian merchant's seemingly tall tales even then. And perhaps such skepticism was deserved, given Rustichello's fulsome prologue asserting that "from the creation of Adam to the present day, no man, whether Pagan, or Saracen, or Christian, or other, of whatever progeny or generation he may have been, ever saw or inquired into so many and such great things as Marco Polo." When Polo was on his deathbed, several nobles of Venice visited him to extract confessions, threatening that this was his final chance to come clean. But Polo, defiant to the last wheeze, said, "I did not tell the half of what I saw."

Neither did we, certainly not to our parents. When Mel and I called home with updates from the Silk Road, we told them the better half of it: the warmth and hospitality of the Turkish people, the delicious meals that rendered scurvy impossible, and the sweeping, moody views across the Black Sea, that liquid frontier always to our left. Maybe Polo also realized that some things are best left unsaid.

The wristwatch alarm buzzed from the tent ceiling, barely audible against the drum of rain. I ignored it while Mel immediately jumped into gear. She pulled on soggy bike shorts and damp long underwear and soaked rain gear, rolled up her sleeping bag, deflated her sleeping pad, stuffed her belongings into panniers, and started boiling water in the tent vestibule, warming her hands over the steam. I still didn't budge, immobilized less by physical tiredness than despair. If Ben had met his limits on the remote, tortuous back roads of Tibet, I'd hit mine on the paved, drizzly, populated Turkish coast. As for Mel, I wasn't sure she had any.

I eyed her from my sleeping bag with a remote astonishment. The distance between where I lay, limp on the tent floor, and where she sat, dressed and ready to bike, seemed intergalactic. Seeing I needed some inspiration, Mel retrieved her journal from its protective plastic bag and read me her list of "Reasons to Go On." So far, she had four:

1. *This first month sucks, but maybe the rest of the trip won't.*
2. *Every other year in your life you can be warm and dry.*
3. *This is a test you don't want to fail.*
4. *There's no alternative.*

I groaned from somewhere deep in my sleeping bag. "Can't we just read and wait out the rain?"

"Drink this," Mel said, handing me a steaming mug of Nescafé. Instant coffee propelled me onto the road. The highway pinched and swerved along the coast, the tarmac a dark river thrashing with cars. To the left the Black Sea kicked up all kinds of colours, the endless nameless shades between seaweed and pearl. I thought I saw a bonito leap, the flash of a bent knife blade and then nothing, and in such moments I could almost recognize the merits of biking this road. But after a few hours of pedalling the caffeine wore off, and any clarity I'd achieved would start shimmying again as though something crucial was loose—a wheel or spoke, possibly a mind.

We wore crinkly plastic shopping bags over our socks and gloves, the only tactic that prevented total soaking until we sweated so much on a climb that we got drenched from the inside out. I didn't enjoy going down again because that's when the chill really set in, and I dreaded stopping for the same reason. Turkish stop signs say *dur*, which is fittingly also the French word for "hard," because the only thing tougher than turning the pedals was *not* turning them. No wonder we resented being pulled over on a daily basis by the Turkish police, or *jandarma*.

"Oh god," sighed Mel as a patrol car squealed to a stop in front of us. "Not again."

Two officers swaggered out of the vehicle to check our passports. When they saw feminine faces peering back at them from androgynous swaddles of Gore-Tex, their tough-guy swagger subtly changed to a preening strut. I huddled next to the open door of the patrol car, relishing the puffs of warmth that escaped, while they consulted our passports—which they held upside down. No matter, it was all a sham, a preamble to what they really wanted: a photo. The officers took turns posing with us, one grinning and giving the thumbs-up, the other snapping photos with his cellphone, then they cheered as we pedalled off into the rain.

A few hours later, we snagged the jandarma for a second time. We were scouting for a place horizontal and hidden enough to camp, which was no easy task on the steep, populated shores of the Black Sea. Finally we spotted a flat-looking field on the far side of the road, but before we could walk over to it, two men with staffs followed by two dogs and a dozen bedraggled sheep crested the hill we stood on. We pretended to admire the view as they shuffled past us, trailing the smell of damp wool. Once they were out of sight, we prepared to dash across the road, but two vans full of jandarma pulled over. A dozen uniformed men bristled around us.

"Where you go?" an officer demanded, his every feature thickened with authority: fists meaty as steaks, arms like oversized kebabs. I tried not to be immediately irritated with him.

"Hindustan," Mel told him. This is Turkish for India, and the truth, if not the answer he was seeking.

"Why?" he demanded, gesturing at our bikes.

It was a reasonable question to which there was no reasonable answer. Why bike the Silk Road? Because it's there, sort of, in a historical and metaphorical sense. Because I wanted to seek out the world's wildness and plumb my own in the process. So far, I was sorry to report, it was roughly as shallow as the oxygenated surface layer of the Black Sea, perhaps because searching for wildness on civilization's oldest superhighway was a flawed premise from the start.

But I said nothing. Mel levelled a hard look at the policeman, her face flecked with mud, her clothes soaked with rain, her legs tied in knots with tiredness—at least if they felt anything like mine.

"Because it's fun," she told him grimly.

The officer raised his eyebrows. "Are you married?"

Our standard answer was yes, of course; we were married to beefy Turkish truck drivers named Osman and Mustafa, who were following us in support vehicles and would probably catch up any second now. When a truck conveniently roared into sight, as they did every few seconds, we'd wave and smile at the driver, and the startled man (it was always a man) would wave back and sometimes even honk his horn, lending credence to our narrative. A Turkish friend had helped us contrive this story, insisting it was our "best only insurance policy." But this stretch of the road in the late afternoon was remarkably quiet, so Mel and I simply pulled off our gloves to display our fake wedding rings.

"How many children?" he demanded next. When we told him *yok* he seemed mildly impressed with our grasp of Turkish. He proceeded to inspect our bikes: pinching the tires, testing the brakes, trying and failing to lift the loaded frames. By now some of the other policemen were discreetly snapping photos of us with their cellphones. The officer asked where we would sleep. I told him *çadir*, Turkish for "tent."

"Terrorists," he said disapprovingly, sweeping his arms to signal the evil skulking in the hills. "No good for the ladies."

Mel and I looked around and saw no evidence of either terrorists or ladies.

"Tonight, you go Samsun," he said, making a cycling motion with his hands. "You stay hotel. Okay?"

Samsun was sixty miles away and it was nearly dark.

"Okay!" we affirmed. "No problem!"

The jandarma drove off and we scuttled across the road. Mud glazed with frost sucked at our feet, blazing a clear trail to our campsite. In the bare field, the Glow-worm's red fabric looked as subtle as a flare gun. The tent was so huge we could park our bikes in the vestibule and still have room to sleep, cook dinner, and host a dance party if we had the energy, but most nights that first month we were too tired to even talk. "Argh," I'd grunt at Mel, and she'd pass me the water. "Mrmph," she'd mumble to me, and I'd swab toothpaste on her toothbrush.

Brushing my teeth outside the tent that night, I spotted two human silhouettes on a distant ridge holding what looked like guns. Was it the jandarma, tracking the terrorists? Or the terrorists themselves? Or the herdsman who passed us earlier, shadowed by wet sheep, carrying walking staffs? That night we slept with a holster of pepper spray stashed between our sleeping bags, just in case.

When some fluke rays of light strayed onto the road a few days later, I swerved into the gutter, convinced a transport truck with its high beams on was bearing down on my bike. Then I realized it was just the sun, that pale asterisk in the sky, referring to a footnote at the bottom of the Black Sea that reads, in very fine print, "Shines hotly in theory."

When I said this to Mel she looked at me like I was nuts, but even she was beginning to lose her mind. At one point the road swerved through a dark, dripping, two-lane tunnel that was nearly three miles long and dangerously narrow. The safest way through it was along the raised ledge to the right side, just wide enough to push our bikes along, but not wide enough to avoid getting lanced by the side mirrors of speeding transport trucks. Luckily we could tell when those vehicles were approaching—and could squeeze against the wall accordingly—because the tunnel would roar as though collapsing under its own weight. Before we set off, I asked Mel to say a little something for the camera. She mumbled about needing to travel the tunnel at a "quick trot." When we finally emerged into relative brightness on the far side, shaken and nearly deaf but intact,

Mel couldn't stop giggling in a slightly unhinged way. "I thought my dying word, captured forever on film, would be 'trot'!"

Finding places to pitch the tent along the busy coast was a challenge. By dusk one evening we still hadn't found anywhere to camp. White flakes silvered the sky and for a second I thought it was snowing, but then I realized that it was ashes wafting from some fire. Garbage burning, I guessed, by the stink of it. Or maybe that smell was the Black Sea flipping over without warning and spewing up hydrogen sulphide the way anoxic basins are prone to do. Apparently you get one sniff of rotten eggs before the chemical obliterates your sense of smell, making it impossible to tell whether you're inhaling more of the deadly miasma. After three weeks in Turkey, I wished highway pollution worked the same way—not to the point of poisoning Mel and me, of course, but enough to shield our smell receptors from the truck fumes. Finally even my kickstand couldn't take it anymore. Unprompted by me, with the bike wheels still spinning, it put its foot down as if to say, *Enough is enough.*

We walked our bikes onto a random family's lawn, my broken kickstand dragging. A woman was yanking weeds from a garden, her face turbulent with wrinkles, her spine curved into a parenthesis. Later we were shocked to learn she was only in her forties. *Merhaba,* "hello," I called out, and she shuffled over with wide eyes. The two of us looked like emptied oysters, our innards digested and slurped out by Japanese whelks, our eyes two neatly drilled boreholes.

"*Kamping?*" I ventured. This is basically the same word in Turkish as in English, only pronounced with a subtle twang I could never quite master. The woman had no clue what I was saying. I tried hand movements suggestive of a tent, sleeping, cooking dinner.

"Let's just set up camp," sighed Mel. "Easier to mime an apology than a request for permission."

We hauled out a wet slump of nylon from the tent bag. Several other people joined the woman to watch us untangle the ropes, click the metal poles together, slide them into a sodden mess of fabric, and abracadabra, tug entropy into an abode. Our audience clapped with quiet appreciation. Only after we'd unloaded our bikes, unfurled our sleeping bags, and primed the stove to boil water did a lean older fellow, whose name we learned was Hasan, admonish us to pack it all up again. It was too cold out here, the ground too hard, he explained, or we thought he explained. His gestures were mystifying, but the gist seemed clear: we were invited to stay with his family.

The walls of Hasan's home were covered in faded linoleum, the furniture draped with prayer beads. The wood stove made such a sauna of the main room that I could understand why he'd deemed our tent unlivable. His niece and two daughters were fashionably dressed, at least relative to Mel and I, who looked shabby and overheated in fleece pants and tops. I was suddenly grateful that my fingernails were still hot pink and hardly even chipped when the youngest girl, Fatma, admired them.

Hasan was a farmer in his sixties, as we gleaned from his niece, who spoke some English. But while farming was Hasan's day job, drama was his true calling. He continued to express himself with bombastic gestures that grew more theatrical as the evening wore on: thigh slaps and raucous hoots at odd times, finger taps on his nose, tugs on his earlobes, each movement blazoned with cryptic meaning. He jointly nicknamed us "Melika" and employed this moniker as often as possible. I caught his wife, her sweet face squeezed in a flowered head scarf, smiling at his antics as she served us shelled hazelnuts and tea the colour of teak.

When she disappeared to prepare dinner, the rest of us turned to watch the news on television. The weather report was the only part that made sense to me, all those cartoon rain clouds covering little yellow suns. After the news we watched some kind of reality drama about a woman marrying a disabled man. The bride wore a creamy white dress, the groom in the wheelchair a natty black suit. Although the circumstances behind the wedding were unclear, it seemed as though outraged people were calling in to contest the union. A smarmy television host arbitrated the calls, his wiry arms gesticulating madly as he said who knows what. The couple held hands and said nothing. Her veiled face was fixed on the ground; his eyes gathered sorrows at the side of the room. I was horrified by the whole thing, but perhaps I misunderstood what was going on.

Hasan's wife reappeared with cabbage rolls and bowls of lettuce drizzled with olive oil and lemon juice. She tore bread into bits and tossed them on the table to be used as utensils. As we ate dinner, Hasan asked—through his niece—about our jobs. Mel said she'd studied capacity building and food security in rural communities. Based on the niece's translation, Hasan took this to mean that Mel was a farmer, like him, and slapped his knee in delight. Then he turned to me. "A lapsed scientist," I told his niece. She looked confused. "A wannabe explorer." Was she baffled because she didn't understand, or because she knew I might as well aspire to be a mastodon? I searched for some other neat descriptor of who I was, rather curious to know the answer myself. Only one label seemed sufficiently vague and quixotic to suit me now.

"A writer?" I offered, and showed Hasan my journal. When he took it and flipped through the pages, I suddenly feared he might read my despairing rants about Turkey's murderous traffic and overpopulated coast, the parabolas of pain that were its roads. Finally he held up my journal and announced, with characteristic flourish, "Jalal al-Din Rumi!" At least that's what I later guessed he said, because all I heard, in the moment, was a string of syllables ending in *Rumi*.

Relief washed over me. If he'd confused my scribblings for poetry, this was proof he couldn't read the journal's contents. I'd been surprised to learn most Turks knew of the thirteenth-century Sufi poet, though I shouldn't have been, because he's something of a national icon. Born in Afghanistan, Rumi's family migrated west to avoid the conquering hordes of Genghis Khan and eventually settled in Konya, a city in central Turkey, where he lived as a wealthy nobleman and scholar for many years until a charismatic desert wanderer named Shams showed up. According to legend, Shams shoved Rumi's treasured books into a fountain, declaring that it was time for Rumi to start living what he'd been reading and talking about for so long. So began an impassioned friendship that inspired the whirling dervish Sufi order and more than seventy thousand lines of verse. I'd been scanning many of them in the tent on my e-reader each night, which I'd loaded with hundreds of books, fiction and non-fiction, though all I seemed to read on it was poetry. After a long, tiring day on Turkish highways, I craved the sort of lyrical intensity this stretch of the Silk Road seemed to lack, and I also craved brevity: a few compressed stanzas were all I could take in. Mel, however, was soldiering through *War and Peace*.

Her stamina in words was surpassed only by her social endurance. At the moment she was showing our hosts photos of her family and life back home. The ladies crooned over Mel's boyfriend, making a gesture like plucking grapes from a bunch on a vine, which appeared to be a sign of approval. I zoned out, drank tea, and thought about how scholar and translator Coleman Barks described the psychic state of Rumi's poems, namely "heartbroken, wandering, wordless, lost, and ecstatic for no reason." I'd copied that sentence into my journal with a nostalgia close to pain: it's how I'd felt biking across the Tibetan Plateau, where every day was tensioned between joy and suffering, heaven and earth. But then again, I wondered, couldn't the same be said for Turkey, if on a less grand scale? Happiness was sipping sugary *çay* next to a wood stove in a tea shop when catatonic with hunger and cold, or the moon spending its silver light over the sea, or total strangers treating us like lost family. Heartbreak, in one of its milder iterations, was how the road—I swear it—always went up. "The only rule is," counsels Rumi, "suffer the pain."

Eventually the women noticed me yawning. Taking our hands, they guided Mel and me into our bedroom for the night. After Hasan had protested about us camping on the cold, hard ground, we were left to sleep in an unheated room on couches stuffed with granite, or some material like it. Mel and I looked at each other and laughed. At least it was dry. I crawled under a thin wool blanket that coughed dust with my every fidget and thought about the reality show couple, wondering how things had worked out for them. The television blared more news through the wall and soon enough I was asleep.

5.

THE COLD WORLD AWAKENS

Lesser Caucasus

*A*ll along the Black Sea I kept confusing clouds massed on the horizon for mountains. When the rain stopped one evening, near Rize, we finally turned a corner that revealed the mirage as real. Serrated peaks flared coral above the city, just for a moment, then the sun sank and the Kaçkar range snuffed out. But in that brief glimpse all was forgiven of Turkey: the rain, the traffic, the chest cold I'd developed and couldn't seem to shake. Michael Ondaatje says the first sentence of every novel—and every travel book, I might argue—should be: "'Trust me, there is order here, very faint, very human.' Meander if you want to get to town." But who wanted to get to town? I wanted to get back to the mountains.

And there they were, many-bladed and growing moonward, just a ways down the Silk Road.

That night I coughed through small talk with the Turkish couple who graciously took us in, my throat scratchy despite the endless cups of tea they poured. Better medicine by far was the hot shower they let me take. When I stepped out of the bathroom, a pair of flowery high-heeled slippers was waiting by the door, which gave an oddly glamorous lurch to my walk despite (or perhaps because of) the fact that they were many sizes too small for me. The slippers belonged to the couple's young daughter, who giggled when she saw me in them. Her parents urged her to practise English with us over dinner. "Hello my name is thank you!" she offered shyly.

Before we left, the family in Rize scribbled another family's name and phone number on a piece of paper, and in this manner Mel and I were passed like batons between generous friends all across Turkey. The challenge was locating our would-be hosts in the next town, for typically they didn't speak English. We stumbled on a fail-safe tactic: upon arriving we'd head to a busy sidewalk and call the host family's number. As soon as someone picked up, we'd hand the cellphone to a random (and now very confused) Turkish person. "*Merhaba?*" the baffled stranger would speak into the receiver, explaining that two random girls on bikes had handed him a cellphone. The host family he was speaking to would realize the stranger was referring to the foreign cyclists they were expecting. They'd explain where they lived to the stranger, who would hang up and direct Mel and me exactly where we needed to go. That's how we arrived, a few days later, at a half-built apartment tower on a steep slope above Borçka.

The brick building looked stacked rather than cemented, as if a kick would send the whole tower tumbling. A middle-aged man and two little girls, maybe eight or nine years old, were waiting at its base. They helped ferry our bikes and bags up crumbling concrete steps, past rusty entrails of rebar, and into a beautifully finished apartment that smelled like fresh-baked bread. The living room contained the usual altars of a television and a portrait of Ataturk, whose icy blue stare seemed to follow me wherever I went in the room. Not that I could move around much, for the floor, couch, and several chairs were packed with grandparents, uncles, aunts, and half a dozen kids, among them an adorable four-year-old with rosy cheeks, ample batting eyelashes, and a demonic soul.

She dumped bowls of oily soup on the carpet and laughed as her mother silently sopped up the mess. She grabbed cushions from the couch and threw them with terrible force and impeccable aim at her siblings. She kicked her frail grandfather in the shin and laughed more as the old man howled in pain. Her family clucked disapprovingly at these antics but didn't intervene; everyone seemed half in awe of her, half terrified. "Ahh, comic," the grandma smiled weakly but indulgently, as if to say kids will be kids. And psychopaths will be psychopaths, I thought, narrowly avoiding getting kicked in the shin myself as the child apparently read my mind and decided to punish me.

That night Mel and I piled panniers against the bedroom door to stop the comic from sneaking in. At least the little girls whose room we shared were sweet, if also riveted by our every move: the way we brushed our teeth, the long johns we wore, the fact that we stared mutely at inanimate bricks for an hour before going to sleep. This absorption in reading seemed to puzzle them most of all. I'd rarely seen books in Turkish households, other than school textbooks, and I wondered what happened to would-be readers and dreamers who grew up in small towns here, especially with parents who emphatically vowed they'd never let their daughters bike across Turkey. Lying there half-asleep, I felt an overwhelming love for my own parents, who'd encouraged all kinds of exploration when I was a kid, in words and the world, though perhaps they regretted raising me to be a little too fearless. A few months after I'd told them about my plans to bike along China's Silk Road for the first time, Mel and I were packed and ready to go when, forty-eight hours before our flight, she went for a last swim in the lake at her family's cottage and was run over by a motorboat. She managed to avoid the propeller but the boat rammed her thigh, generating a deep muscle contusion that would require at least a month of physiotherapy to regain mobility. As far as biking in China that summer was concerned, Mel was a bust.

I'd received the news in North Carolina and hesitated over how to proceed. After all, I didn't know how to speak Mandarin or Uyghur or Tibetan (neither did Mel, for that matter). I didn't know, despite my best intentions to learn, how to fix a flat tire. And now, quite suddenly, I didn't have an expedition partner. It seemed clear what I had to do.

"I'm going anyway," I told my parents over the phone.

"You are not."

"I'll be fine!"

"Forget about it!"

"Love you, bye, leaving for the airport now!"

When I flew only as far as California, instead of Beijing, and started pedalling back to North Carolina, instead of along the Silk Road, my parents' relief about me biking solo and coast-to-coast across America was directly proportional to the relative risk of what I'd proposed to do. This proved a canny lesson in future expedition planning. "Mom, Dad, I'm going to Mars," I'd announced gravely after my meeting with Zuber at MIT, explaining with regret that it would probably be a one-way ticket. Then, amid their loving protests, I'd graciously relented and proposed the real plan. "Okay, okay, fine . . . I'll just finish biking the Silk Road, from the Caucasus to Kashmir, skirting the Afghan border and sneaking across Tibet again along the way."

I glanced over at the girls, wishing I could share these expedition planning tactics, spur them into a life of experiment and adventure, but I didn't have the language and they were already fast asleep.

The world, seemingly overnight, went wild with multiplied cold. Mountains loomed steeply on all sides of the road, and creeks cut the air with their freshness as we passed. Freezing rain on the Black Sea had turned to falling snow higher up, but the road was plowed to a thin layer of white. Though traffic was sparse, it wasn't nonexistent, and at one point a lone sedan slowed to a crawl as it passed. I saw curious faces pressed flat against the windows but didn't think anything of it until we crested a hill and saw the car parked on the top. Waiting on the road shoulder was a tall spindly man in a tweed jacket who sported a greasy comb-over, so that his thinning hair was rendered eerily motionless in the wind. He aimed a camcorder our way as we approached. "Where are you from?" he quizzed us. "Where are you going? Are you cold?"

"Canada, Hindustan, not cold, though if we stay still we will be. Goodbye!"

We didn't get much farther. The snow grew so deep and the road so steep that Mel could barely move. "It's not fair," she complained, her wheels whirling in place like dervishes. "I'm pedalling twice as hard as you and going half as fast." I managed to maintain some forward momentum thanks to the valuable skills I'd learned in graduate school at MIT, namely biking recklessly fast on slippery terrain. Figuring it wouldn't be wise to exult about how much I loved this technical riding, the way it forced a total focus, I suggested we take a break at the nearest bakery. When the lady running it brought over our order of baklava and Nescafé, she said the road to Ardahan was *kapali*, closed. Mel didn't seem disappointed. "More Nescafé?" she suggested brightly, settling in with her book.

A few hours later the road opened, but only to four-wheel-drive minibuses with chains on their tires. The jandarma had already forced us to hitch a ride over a previous high pass, because of an earlier blizzard, which meant I was doubly reluctant to load my bike onto the roof of another vehicle. The bus struggled up and onto the Kars Plateau, which like Tibet was buckled skyward by colliding land masses, in this case Eurasia and Arabia. The resulting sweep of plains and mountains lies more than five thousand feet above sea level and borders Armenia, Iran, and Azerbaijan. Although much of Turkey enjoys a temperate climate, *kar* is Turkish for snow and the plateau spells it out in the plural—at least when we arrived in February. As the bus sped past horizons swept clean by wind and snow, the kind of landscape on which light falls in huge cold slabs, I silently fumed with regret. If we were going to travel parts of the Silk Road by car, why didn't we skip the murderous traffic and freezing rain on the Black Sea?

"Thank *god* we're not biking this stretch," muttered Mel.

We reached Ardahan in the late afternoon and ducked into a restaurant. While eating lentil soup, we happened to glance at the television just in time to see ourselves broadcast on the news. Cue the spindly man in tweed, speaking with commanding authority into the camera. Cue us slowly cresting the hill, muttering a few words between panted breaths, and disappearing into white oblivion. Best of all, this particular channel gave news stories the full Turkish soap opera treatment: slow-motion clips played on repeat to a musical score better suited to crime dramas than reportage.

The story about us was no exception, though Mel and I biked away at such a glacial skid that we supplied our own slow motion, no special effects necessary. Once we disappeared after an awkwardly long time, the newsman turned the camera back on himself and resumed his authoritative commentary.

"What on earth is he saying?" Mel marvelled. "We barely spoke to the guy!"

I glanced around the restaurant to see if anyone had connected the news with us, but perhaps we were hard to recognize without our scarves, hats, and helmets. People went on sipping tea, their eyes fixed on the screen, oblivious to the "celebrities" in their midst, which was fine by us. We bundled back into our warm layers and wheeled our bikes into the lung-crackling cold.

When I woke the next morning the tent ceiling was constellated with frost. All the stars seemed alien, ungathered, and for a moment I felt unsure what planet I was on, the sky above suspiciously crimson. Then I spotted an earthly landmark in the tent's laundry line, where two pairs of wool socks and my watch drooped stiffly. I sat up to check the time and accidentally brushed the tent wall, sending the visible universe into supernova. Frost flaked off the ceiling, the fabric of space-time buckled and creased, frozen socks drop-kicked my lap. It was eight in the morning.

"Are you awake?" I whispered to Mel, eager to hit the road.

"No," she whispered back, flakes of ice settling on her lashes.

I crawled out of my sleeping bag in my long underwear and dressed in all the clothes I carried: fleece pants, a fleece top, snow pants and a matching shell, a down jacket patched in a dozen places with duct tape. That jacket was a map of all the cold, lonely places that had mugged me with their beauty: the Tibetan Plateau on our first bike trip, Norway on a ski traverse during a break at Oxford,

Kashmir after I quit MIT and two friends and I followed in Fanny Bullock Workman's footsteps up Pinnacle Peak, the 22,000-foot Himalayan mountain on which she set the women's world altitude record in 1906. The colder and harsher the terrain, the more I seemed to come alive in it, but Mel lacked my enthusiasm for deep-freezes. Her sleeping bag showed no signs of movement, so I read her some inspiring lines of poetry I'd copied into my journal. "What is the colour of wisdom?" wrote the poet Evan S. Connell. "It must have the colour of snow." Mel groaned from somewhere deep in her sleeping bag. Nescafé, stat.

I lit the camp stove in the front vestibule with the door cracked open for ventilation. Our lone titanium pot was welded with burnt noodles from last night's dinner, but I filled it anyway with water from a bottle I'd cuddled with all night to keep from freezing. The resulting coffee tasted like cinders, and not even gobs of peanut butter could override the burnt flavour of the oatmeal. "Let's just save this for later," I said to Mel as the gluey mass froze in our mugs, explaining that at times polar explorers were reduced to chewing their leather boots for calories. For some reason Mel didn't find this reassuring.

Hours after waking, we finally crawled out of the tent and jumped in place to get warm. Packing up required the dexterity of bare hands, and our skin stuck to all the metal that made up our life: the stove, tent poles, bikes. We rolled the frozen-stiff fabric of the tent into its frozen-stiff bag, then dragged our loaded bikes through deep snowdrifts back to the road.

The British Antarctic explorer Apsley Cherry-Garrard claimed that "polar exploration is at once the cleanest and most isolated way of having a bad time which has been devised." Winter bike trips in Turkey might be a close second. Yet when traffic was scarce, when it wasn't raining, when mountains set the mood of the road,

when I wasn't aching and hungry and miserable—and even sometimes when I was—biking was the cleanest and most isolated bliss
I knew. Even Mel seemed to be enjoying herself, for every so often
she would stop and turn cartwheels on the road ahead of me. I
kept stopping as well, but to snap photos: of the air scrubbed clean
and blue with wind, of the mountains smoothed over with snow,
of a landscape indifferent to my admiration, and all the more
compelling for it. More sky than earth. More wind than world. No
wonder Kars sent me soaring.

In that respect, I wasn't alone. Millions of birds wing it over
the plateau every year, travelling from western Siberia and the
Middle East to southern Africa and back, though we were sadly
months too late (or too early) to see the sky storming with feathers.
The migratory highway through Kars is especially popular with
birds of prey: hawks, eagles, vultures, and falcons that skip like
stones from thermal to thermal—vortices of warming, rising air
that form as different surfaces absorb different amounts of sunlight. Raptors soar effortlessly up one thermal, then glide down to
meet the next one forming, and in this way travel thousands of
miles without flapping a wing. I've often wished bicycles could be
powered on the same principle.

Maybe all flight begins with the envy of birds. In the mid-
nineteenth century, the young Otto Lilienthal longed to soar with
the storks that ruled the skies in his German hometown. Not that
he dared voice that ambition: back then the idea of building a
flying machine was akin to designing a perpetual motion machine
or transmuting lead into gold—the typical preoccupations of
cranks and dreamers. With the help of his brother, Gustav, Otto
apprenticed himself to wind and wings, often working at night to
avoid being seen by gossipy neighbours. By trial and error he
gleaned that flight is easiest when you launch *against* the wind,

rather than with it, because the faster air moves against a set of wings, the more lift is generated, meaning headwinds don't impede flight so much as give it a turbo boost. He began testing gliders by day, when the winds were strongest, and crowds gathered to jeer. As the gliders flew a little farther each time—thirty feet, three hundred feet, a quarter mile—they began to cheer. Over the next decade, thanks to Otto, the notion of human flight went from a foolish lark to a serious science. "The time has passed," he proudly declared, "when every person harbouring thoughts of aerial flight can at once be pronounced a charlatan."

I thought back to the photographs of Lilienthal's gliders I'd looked through at Oxford, where I'd been struck by how birdlike they were, how plainly Icarian in inspiration, with wings built from feathers and sticks as if pure mimicry could provide sufficient lift. The design of flying machines gradually moved away from avian whimsy toward a more clean-edged efficiency—not because bird wings can't do flight best, but because we've failed to mimic their flapping in more than a century of trying. Even now, as people snore on transatlantic flights and yawn at yet another rocket launch to the International Space Station, bird wings represent a union of mechanical efficiency, fuel economy, and metaphorical grace that no human invention has ever matched. Except, perhaps, the modern bicycle.

The clumsy ancestor of the sleek machine I rode appeared on the streets of Paris in 1876. Lacking pedals, drive trains, and pneumatic tires, these two-wheeled "dandy-horses" could only be propelled by kicking the ground, earning them the name *velocipede*, from the Latin for "fast feet." Because velocipedes were small and less visible than a horse or a carriage, they gave their riders the "comical appearance of flying through the air," as *The New York Times* rather mockingly reported. But just two decades later, right about the time Lilienthal's glider flights were making headlines, velocipedes had evolved into the "safety bicycle," a machine more akin to our modern counterparts and whose relatively cheap price tag, comfortable ride, and ease of handling effectively granted humans wings, including Fanny Bullock Workman and her husband, who rode them across Europe and India. By 1896, *The Aeronautical Journal* was describing the similarities between cycling and flight not in jest, but in earnest: "It was not uncommon for the cyclist, in the first flash of enthusiasm which quickly follows the unpleasantness of taming the steel steed, to remark, 'Wheeling is just like flying!'"

I said as much to Mel when I caught up with her in Kars, but she failed to agree. "I didn't sign up for this," she moaned, jogging in place. "I can't feel my fingers, my toes. I can't even muster the *memory* of what fingers and toes once felt like." It dawned on me that her cartwheels hadn't been expressions of joy, but attempts to centrifuge blood to her extremities. "This isn't safe," she continued. "We're in the middle of nowhere and I'm freezing and *I can't do this anymore.*"

She sounded close to tears. I couldn't verify this because Mel's face was concealed behind a balaclava and sunglasses, the kind of glamorous, oversized frames that celebrities wore to hide identities and hangovers. Even way out here, on the frozen Silk Road, I thought meanly, a part of her still insisted on looking cool.

For a spell as teenagers, Mel and I hadn't been friends at all. We went from being inseparable in elementary school to nodding curtly at each other in our small-town high school, where home-made magic tricks quickly lost their social currency. I dreamed of becoming an astronaut and going to Mars, the last frontier left for a wannabe explorer. Mel, by contrast, was popular. She wore makeup to school and partied on weekends, while I was the kind of student who completed school assignments weeks ahead of time and suffered paranoid delusions on laughing gas at the dentist, convinced I'd be disqualified from NASA for taking drugs, even nitrous oxide for the filling of cavities.

We barely talked for years. I suppose we both felt hemmed in by our small-town high school, with its narrow hallways and narrower minds, but we responded to these constraints in clashing ways: Mel strove to excel within them, to belong and be cool, while I was obsessed with elsewhere, with escape. I couldn't feign an ironic detachment from the world if I tried. It was a relief when my family moved again and I switched into a larger and less cliquish high school, where I poured my energies into riding horses, learning to skateboard, and plotting my launch into space. Which perhaps explains why, on the Kars Plateau, a petty, regressive part of me thought that if Mel had spent less time in high school being popular, and more time studying the polar expedition narratives of Cherry-Garrard and Shackleton and Nansen, she'd have a richer understanding of what constituted hardship and extremity. This wasn't suffering; this was adventure!

"We're hardly in the middle of nowhere, Mel," I began unsympathetically. "I wish! But no, we're on a paved road, we have a cellphone, and there's even a gas station just ahead. It's a joke, we can bail in milliseconds if we need to. There's no safer place to be cold in the world—and it's not even that cold!"

Some pep talk. Mel didn't say anything, not even to justifiably point out that she'd been the one to coax *me* out of my sleeping bag on the Black Sea. She just got on her bike and rode away, only to slip on black ice and crash into the pavement. When she crashed a second time farther down the road, I felt sickened by my previous smugness and accelerated toward the gas station, willing it to be open. By the time Mel walked her bike over I'd bought two packages of cookies and three chocolate bars, and in silence we ate them really fast. The lone gas station attendant fetched us hot water to mix with Nescafé, and even went so far as to scrape the frozen gruel from our mugs, and in these simple gestures it seemed possible to rebuild the world.

~

Do grudges ever go away, or do they only go dormant, like black seeds slinking just below the surface awaiting ideal conditions to sprout? I wanted to believe in a world without borders, which meant believing in hearts and minds without them, too, and this wasn't easy when I looked into my own. Or when I looked around the Kars Plateau, a region variously known as eastern Anatolia or western Armenia, depending on who you asked at the turn of the twentieth century.

Back then, this region was violently contested between Turkish, Armenian, and Russian forces. Much of the Armenian population in Kars was killed in what the Turks still refuse to call genocide, and the bloodshed didn't cease until the Treaty of Kars in 1921 ceded most of eastern Anatolia/western Armenia to Turkey, including Ararat, the Armenians' sacred mountain. Once perceived as the highest point in the Christian world (though it barely scrapes the bottom of the Tibetan Plateau), Ararat is where Noah's ark supposedly struck dry land after the deluge—at least as far as we can tell, given Genesis lacks GPS coordinates. The Turks renamed the peak Agri Dagi, or "mountain of pain," in what seemed a pointed nomenclatural demotion from sacred to profane. With the fall of the USSR in 1991, Armenia regained independence, but not its revered mountain, which loomed out of reach across the border with Turkey somewhere down the road we were travelling.

"Is that Ararat?" asked Mel, pointing at a snow-capped summit from the car that had picked us up not far from the gas station.

"No, no, not yet," growled Onder from behind the steering wheel, not because he was angry but because he always spoke that way. Dishevelled and rotund, the thirty-something nature conservationist reminded me of a bear prematurely woken from hibernation. Perhaps this wasn't far from the truth: while he'd been expecting our call, it came earlier than anticipated, namely from the gas station where we stopped for snacks. Mel and I biked on a little farther to the turnoff to Georgia, where Onder met us and would drop us off a week or so later, to continue biking to Tbilisi from exactly where we'd left off. In the meantime Onder had offered to show us the conservation projects run by his employer, KuzeyDoga, a local non-profit working in the Turkish borderlands, including at the base of Ararat, or Agri, depending on where you stand.

"How about that one?" I asked when another large peak loomed into view.

"Trust me, you will know Agri when you see it," said Alkim, cracking sunflower seeds between his teeth. He had jokingly introduced himself as a filmmaker "famous in northeastern Turkey and parts of Iran," which sent Onder into high-pitched giggles. They were old friends, working together on a film about KuzeyDoga, and Alkim looked kind of like Onder, only stretched—taller, thinner, smoother—and doused with considerably more cologne. When he tried to spit sunflower seed shells out the car window, they blew back and got stuck in his glasses.

We rounded a bend and I realized Alkim was right: Agri was so obviously Agri, or was it Ararat? Perhaps I'm biased in believing all mountains holy, whatever their names, but this one in particular, with its towering twin peaks, looked particularly divine. It was less an upheaval of rock than a cold slump of stars. The usual minaret towered above the village of Aralik, but its relative prominence was beggared by the peak behind it. The village's buildings looked excavated from, rather than built with, the basalt scattered like birdseed around the volcano's base. Children herded livestock through the streets, swatting at the bony rumps of cattle with reeds from the marshes next to town—the last dregs from the deluge, perhaps, supplemented by a steady drip of glacial meltwater.

Similar wetlands flank the sacred peak on all sides, spilling across the borders of Turkey, Armenia, Iran, and Azerbaijan. Because the marshes sit at such a low elevation, down below the edge of the Kars Plateau, they remain an unfrozen oasis for water birds year-round. As we admired a pair of herons wading through one marsh on stilt legs, a herdsman roared up on a shiny blue motorcycle. He spoke with Onder for a while, gesticulating at the wetlands, the fields, the village. When the man sped off, blurring the horizon with dust, we asked Onder what he'd said. "He wants to drain the wetlands, replace them with fields for growing crops."

While the lifeless peak of Agri is a national park, the marshes at its feet lack protected status, which seems a little like valuing the ark more than the biodiversity it carried. KuzeyDoga was trying to persuade the Turkish government to designate the marshes under Ramsar, an international treaty for the conservation and sustainable use of wetlands, which would protect the habitat while helping to promote birdwatching in the area, a potential source of tourism income for villagers. "This is the best part of you being here," Onder said. "It shows them that this place, their home, is valuable as more than grazing land."

I hoped he was right, but I wasn't so sure. One Turkish word for foreigner is *gavur*, which historically meant "infidel." I was also slightly uncomfortable as an ambassador for ecotourism, which encourages people to see dollar signs when they look at a bird or a wetland, though I suppose they already saw money in the marsh in the form of grazing land for animals, or income from growing pulses or grains. I didn't blame them, not a bit, especially when I breezed in, swooned over the birds and mountains, and would soon breeze out again. Putting a price tag on wilderness can pay off, especially in places like Aralik, where the locals need support and the wetlands need protection and ecotourism could enable both. But I worried that something crucial gets lost in such transactions, namely a recognition that the world has value and meaning beyond its usefulness to us.

We walked back to the car, now surrounded by shaggy grey donkeys nibbling on grass. They didn't blink when we opened the doors and slammed them shut. Onder swerved to avoid first the donkeys and then the marshes that the Turkish government had so far refused to protect, perhaps because Armenia was enthusiastic about the idea. KuzeyDoga was trying to get the locals on board in hopes the government might come around, but nature conservation doesn't translate well into Turkish. Even Onder's grandma was still convinced he was unemployed, and her confusion was understandable given his job ran the gamut from collaring wolves to petitioning governments to peeling dead animals off the road. "Remember!" reminded a note taped to the car's vanity mirror. "Do roadkill surveys while driving out of town."

Mel rolled down her window and stuck her head out, savouring the hint of spring in the air now that we were past the donkeys. Ararat seemed to grow bigger behind us, gaining heft and stature with distance, and the wind rustled in the reeds. Out another window I noticed an enormous trident on the horizon, its three stubby prongs spewing smoke or steam, I couldn't quite tell.

"Oh, that's Metsamor," Onder clarified cheerfully. "The next Chernobyl."

This Soviet nuclear reactor was built on shaky ground. After a 1988 earthquake killed 25,000 Armenians, with its epicentre less than sixty miles from Metsamor, the plant was closed for safety reasons. But when the Soviet Union dissolved just a few years later, the newly independent republic of Armenia found itself desperate for cheap power, in part because Turkey and Azerbaijan, its nemesis neighbours, had deliberately routed a natural gas pipeline to circumvent the country. So the Armenian government resurrected the antiquated reactor, which fumed defiantly only six miles from Turkey and not much farther from Azerbaijan. *If I'm going down*, Metsamor seemed to menace on the border, *I'm taking you all with me.*

"Nationality is babyishness for the most part," said Ralph Waldo Emerson, and for the most part I agreed. The more I learned about the South Caucasus, with its closed borders and warring enclaves, the more the place seemed like a playground game of capture-the-flag turned vicious, all in the dubious name of nationalism. And yet political frontiers, while sometimes solid as brick, are finally only as strong as shared belief—the flag-waving faith that the name "Turkey," say, or "Armenia," represents some kind of genuine, immaculate sovereignty, etched out and inviolable. But when Polo travelled through the South Caucasus in the thirteenth century, he visited Silk Road territories long since vanished or metamorphosed, such as Lesser and Greater Hermenia, Turcomania, Georgiana, and Zorzania. "Names are only the guests of reality," the Chinese sage Hsu Yu noted in 2300 BCE, suggesting that borders are little more than collective myths—fictions that a certain number of people, for a certain period of time, believe are fact.

The ground seemed no less shaky at Ani the next day. Once the bustling capital of Armenia, this former Silk Road metropolis is now mostly ruins on the modern edge of Turkey. Over the past thousand years the city had been sacked by Turks, Georgians, and Mongols, and what they left standing was devastated by an earthquake not long after Marco Polo went to China. The gaunt remains of Ani's cathedrals and mosques suggest openness, but behind them is the hermetically sealed border between modern Turkey and Armenia. Demarcating this frontier is a river, the Akhurian, flowing next to Ani through Arpaçay Canyon, which cuts like a dark scar across a plateau smooth-skinned with snow. Both edges of the canyon are a military buffer zone, strung with barbed wire and regularly patrolled by armed forces. People have been banned from the inside of the canyon for nearly two decades.

For years KuzeyDoga had petitioned the Turkish armed forces for permission to document the diversity of birdlife in Arpaçay. After consent was finally granted, a biologist with the NGO surveyed the length of the canyon, listening for songs and scanning for nests. In the process he discovered a half-dozen aeries of Egyptian vultures. This endangered species of raptor had found a rare swath of undisturbed land in the canyon, ideal breeding grounds between the barbed wires. There is no explaining borders to the birds, but they know a safe haven when they see one.

Staring at the canyon through a window in the mosque of Minuchihr, I was torn between wanting to applaud this oasis of wildness and despairing of the strife that created it. Nature is typically the victim of our blunt and inflexible borders, with barbed wire and brick walls fragmenting ecosystems into useless bits and stopping the movement of migratory species that need to roam as widely as wind. Yet here an endangered species wore the border like a bulletproof vest, finding asylum between the walls our conflicts create.

The canyon reminded me of another frontier I'd visited the year before. After attending a conference in Seoul, I'd signed up for a cheap day tour to the demilitarized zone, or DMZ, a belt of formerly cultivated land on the contested waist of the Korean peninsula. Two miles wide and more than a hundred long, the zone is fortified by steel walls topped with barbed wire, and for sixty years people have been forbidden to enter. During this time the farmed land slowly went feral, cultivated fields sprouted unruly forests, wild cranes flocked to wetlands no longer drained for irrigation, and Asian black bears, leopards, water deer, and other rare species flourished. A war-torn borderland became, in effect, the most fiercely guarded wildlife sanctuary on the planet. Could the same thing happen for Siachen? I'd read about the Korean DMZ at Oxford, for my research on scientific peacekeeping, and I'd wanted to see its resurrection for myself.

"On a clear day like today," the tour guide promised as I boarded the bus in Seoul, "you'll see right into North Korea." But when we arrived at the first stop hours later, the sky was as hazy as gauze over a wound. A Ferris wheel and merry-go-round whirled to deafeningly cheerful music. Restaurants with jaunty names like A Walk in the Clouds and Popeyes Louisiana Kitchen advertised the chance to dine in view of the DMZ. Tourists crowded into gift shops that sold T-shirts, key chains, shot glasses, and other mementoes of the military divide. The whole place had the feel of people laughing at a funeral.

The final stop of the tour was a tower overlooking the southern fringe of the DMZ. The haze had cleared by then, and the sky was the colour of a fading bruise, the pale blue of purest flame or glacial ice. For the equivalent of fifty cents I bought a peek into the DMZ through a spotting scope. Between the walls I saw a gnarled forest with pines, firs, poplars, and willows packed tight after a half-century of unchecked growth. I saw two herons tussling in a wetland, a breeze restless in the grass. In other words, to my shock, I saw wilderness staring back at me down the barrel of a cocked and loaded border.

In every respect Ani was less tense than the Korean DMZ, more abandoned than actively contested, with no souvenir hawkers in sight—just Alkim filming Mel as she gamely pretended to explore the same ruins over and over again. "One more time with feeling!" he requested for at least the sixth take. I made sure to stay out of sight. As the sun blinked cold and low over the mountains, the "city of 1001 churches" caught light the way I wished history would: the crumble and decay illuminated, some foundations still solid, graffiti aged gracefully to art. Walking past a cathedral built a millennium ago, I thought I heard a door slam somewhere deep underground, and ruins thick with dust stirred all answers into motion. Certain places, wrote Jorge Luis Borges, "try to tell us something, or have said something we should not have missed, or are about to say something." What Ani seemed to say was that no story, no wall, has only two sides. All definitions blur, all borders wander, and the longer I stared, the more I swear I saw them move.

When the sun began to set, Onder, Alkim, and Mel made their way back to the car. I scanned the canyon one last time, hoping to spot vulture nests, but what caught my eye were the broken halves of a bridge, that reached toward but didn't quite touch each other above the Akhurian. In the heyday of the Silk Road, Marco Polo might have strolled across that very bridge. The Workmans could've biked over it a century ago. Now only birds could cross the river, where the sides of the bridge met in their reflection in the water.

The next morning Onder dropped us off at the turnoff to Georgia. Alkim filmed Mel and me as we hit the road to Tbilisi, crossing a part of the Caucasus that was once so densely forested, according to Strabo, you could walk all the way to the city without the sun touching your head. The sun didn't once hit our heads either, but only because the cold sky was plated with clouds. Intensive clear-cutting had long ravaged the Caucasian forests of lore, at least in eastern Turkey, where only a few remnant stands of fir, Oriental spruce, Pontic oak and Medwedew's birch survived.

What does it mean to find a broken landscape beautiful? I knew Kars was a tamed relic of its former wildness, but it looked nothing short of sublime that morning. Of course *sublime* is by definition beauty with an edge to it, though usually that edge is some element of danger that a landscape poses to people, not what's left of a place after humans have pared it away. As with the vultures living between the lines at Ani, I wasn't sure whether the beauty of Kars was a solace or a call for despair. Maybe it was a call to prayer, or at least that's what we heard a few hours later, signalling the existence of a town long before we saw it.

We decided to stop in Damal and tackle the pass to Georgia the next day, because the road began climbing just beyond town and there wasn't enough daylight left to make it over. When we inquired at the *çay* salon about accommodations, a thin man with a drooping face offered to let us stay upstairs. He showed us to an empty room whose concrete walls matched the concrete floor that showed through the worn spots of a musty grey carpet. "*Chok güzel,*" said Mel, relieved to have any form of shelter on this chilly night that wasn't a tent. We agreed on a price and the tea house owner left us, explaining with hand motions that he would lock the door and fetch us at seven a.m.

I was just drifting to sleep when I heard a commotion outside. The shouting made it sound like a mob was bent on avenging some small-town evil, and I felt relieved that the door was locked. Then I heard our names. Mel and I went into the front room, which had a balcony, and shone our headlamps down on a street packed with people. Among them, waving their arms frantically, were Onder and Alkim.

"Girls!" they shouted. "Hang on! Don't worry! We will get you out!"

"What are you talking about?" Mel shouted back.

Only later did we piece together the full story: Onder and Alkim had driven up the road to check on us, worried we'd freeze in the tent. When they reached Damal without seeing our camp, they figured we must be staying in town, so they inquired after us in a local shop. "Oh, the foreign girls? They are locked upstairs in the tea house," the shopkeeper told them. Panicked about our apparent imprisonment, Onder and Alkim began shouting outside the tea house to get our attention, prompting half the town to show up to see what the fuss was about. The jandarma showed up as well and ordered the tea house owner to release us. The poor old man came stomping up the stairs, muttering "Why me, why me," under his breath.

Once outside, we managed to convince our would-be rescuers that all was well. No, the tea house owner hadn't locked us up against our will. Yes, we were perfectly safe and warm and happy. Eventually the crowd dispersed, with people looking disappointed that such a promising drama had ended so meekly. As one did in such situations—really any situation in Turkey—Alkim, Onder, Mel, and I decided to have tea. Given that the tea house was closed at this late hour, we trooped across the road to the shop, where the shopkeeper seemed completely delighted by the trouble he'd caused. He poured steaming glasses of tea for everyone, including the tea house owner, who had followed us over, still muttering his mantra of woe.

"He's really upset," I whispered to Mel. She nodded, rueful.

Alkim overheard me. "Upset? What do you mean?"

I told him what the tea house owner was repeating over and over.

"It's not English!" Alkim laughed. "It's nothing, gibberish, like ay-yi-ay-yi or la-de-da."

He explained our confusion to the shopkeeper, who thought it was funny, and to the tea house owner, who seemed pleased by our sympathetic reading of his muttering. At least he smiled as he walked us back to the room and locked us in again.

We were released as promised at seven a.m. It had snowed all night and was snowing still. Mel and I climbed down the stairs and went into the tea house, where our bikes were stashed. Already the room was steamy and packed with men, one of whom kindly brought us tea as we struggled to repair Mel's derailleur. Her bike would no longer switch into the easiest or "granny" gear, which was suboptimal given the highest pass of the trip so far awaited us beyond town.

The man who brought us tea inquired where we were going, and Mel told him, "Gurgistan," Turkish for Georgia. A murmur of disbelief travelled through the tea house. "No, miss, car!" another man said. I protested the idea until I realized he meant *kar*, the Turkish word for snow. The shopkeeper from the night before suggested we hitch a ride with him; he was driving to Georgia that morning. But I was determined to pedal as long as our wheels could grip the road, and even if they couldn't. Thanks but no thanks, we told him, prompting all the men in the tea house to laugh and cluck their tongues.

When we wheeled the bikes outside, though, I worried the men might be right. I couldn't see a road, just a general blankness where the road should have been: white clouds, white snow, and no distinction between them. Mel gamely toiled up the frozen pass in her second-to-lowest gear. I rode behind her, feeling equal parts guilty and grateful that my own granny gear was working. The earth and sky blurred together the way they do in certain moments of flight, and also certain moments of freefall, and for the moment I couldn't tell which was which. Every pedal stroke took it on trust that the world, or something like it, still existed beneath our wheels. And yet we were having a grand time for no good reason, breathing clouds in, breathing clouds out, muscling toward the shared goal of the summit.

In some ways our friendship existed best in motion. Mel and I had reconnected during university, when out of the blue she flew to Chapel Hill to visit me over spring break. Perhaps it was the "y'alls" salting my speech, or the fact that we'd barely seen each other since high school, but she almost didn't recognize me when she landed in the South. "What planet are you from?" she said, smiling as if we'd been best friends forever.

"The red one," I told her, though after wearing a spacesuit in Utah I wasn't so sure. Nor was I sure what to make of Mel's visit, but in general I was too tired to question anything. The day before I'd run the Myrtle Beach Marathon on a whim and untrained legs. I'd signed up to run the half-marathon as part of a charity fundraiser, but when I reached mile thirteen and felt good, I kept going. Every step beyond that distance set a new record for the farthest I'd run in one go, and this exhilarating fact kept me shuffling to the finish line. In the surreal, trance-like state of that run, a tired beach town was briefly rendered mythic: Myrtle Beach seemed elsewhere, otherworldly, proof of the transcendent power of slogging. Only when I stopped did I notice the beaches bordered by sagging motels, the billboards advertising boiled peanuts and Jesus in the same bold fonts and neon lights—and the screaming agony of my leg muscles.

I was effectively crippled in the aftermath. So when Mel casually suggested we run another marathon together, one night over dinner in Chapel Hill, I remember looking at her like she was insane. Then of course I said yes. That fall we began training in our separate countries with characteristic zeal, which meant I soon got shin splints and Mel sprained her hip. But we set off across New York City that November anyway, guzzling water from Styrofoam cups at checkpoints and basically pretending high school had never happened.

Cycling the Silk Road seemed to be the next logical step. Although Mel's swimming accident meant I took off alone from California instead, even the rural back roads of America seemed alien and extreme. Mile by mile I pedalled past the slack-eyed casinos and prim army bases of Nevada, where military jets ripped open the sky and vultures sutured it shut; through the twisted red canyons miming Mars in Utah; over the continental divide of Colorado; and across the picked-clean plains of Kansas, where one dull, rainy morning Mel met me with her bicycle, still bruised but ready to roll.

With enough instant noodles for dinner, and packaged pickles for snacks, there was nothing the two of us couldn't do that summer: sixty miles, ninety miles, a hundred and twenty in a day! Broken spokes, flat tires, we could fix it all! Other than a single day off in Virginia—when we blew our budget on an all-you-can-eat pancake breakfast and felt sick the rest of the day—the two of us pedalled non-stop for a month to reach the edge of the continent. Where the road ended, in Swan Quarter, North Carolina, we boarded a ferry to the Outer Banks, our legs twitching from the strangeness of moving without turning pedals.

Mel and I didn't set any records when we ceremoniously dipped our bikes in the Atlantic, and no crowds cheered. We simply swam in the ocean for what felt like days, and I remember my euphoria at completing the journey blending with a searing pain where the salt water stung my saddle sores. What salvaged our friendship was a shared knack for slogging past checkpoints, past reason and restraint, past the past itself, all the way to a point of transcendent stupor that left our teenage grudges far behind us, as if somewhere along the way we'd arrived in a new land.

We reached the top of the pass far sooner than expected. Suddenly a sign loomed from the blankness announcing "Ilgar Dagi Racigi, 2550m," but this surprised me less than the wolf that loped out from behind it. We soon realized it was just a dog, lanky and grey with long iced lashes. Since she was sweet and Turkish, we dubbed her Baklava. Mel scratched her belly and I fed her trail mix until we decided to continue down the pass. When we pedalled away, Baklava followed.

The three of us coasted for ten miles, the kind of descent cyclists (and probably dogs) adore so long as they don't have to go back up. Tears issued from the press of speed into my face, as though velocity were a deep emotion. I squeezed the brake levers as hard as my frozen hands could, but grit and rain on the Black Sea coast had worn the pads down to metal, so that they practically threw sparks. As the elevation dropped, the temperature climbed, and long runnels of meltwater braided the road. Baklava raced at our side, her pink tongue flopping. She wisely stayed out of the way of traffic, which was sparse, though at one point a yellow construction vehicle sped past us only to stop suddenly on the road ahead. A man climbed out holding a rope. "*My* dog," he asserted about Baklava, though she didn't seem so keen on him. She sat on the far side of my bike, pointedly looking the other way.

"Kate," Mel said evenly, reading my mind. "The border's just ahead. We can't take her across."

Of course she was right. We didn't have the necessary papers nor smuggling capacity in our panniers. Plus the last time we'd tried to rescue dogs on bikes, during our ride across America, things hadn't exactly worked out. The two puppies we discovered abandoned in a ditch outside a megachurch in Missouri had whined constantly through the breathing holes of the cardboard boxes we'd strapped to our bikes—a temporary transportation measure until we could buy a bike trailer in the next town. But when we arrived all the stores were closed. It was a Sunday in the American South, where only the police work on the Lord's day of rest. Defeated, we surrendered the puppies at the sheriff's office, where they promptly peed on the floor and gnawed on wooden desk legs. "Don't you worry," the receptionist said dubiously, swabbing at the yellowy mess with paper towels. "We'll find these sweet things good homes."

"They'll probably be shot," Mel deadpanned as we exited the station.

"Mel!" I'd howled, ready to rescue the puppies all over again.

"Just kidding," she'd said, trying to keep a straight face. "They will be loved by small children."

Such is the joy and heartbreak of life on the open road: puppies, like nations, come and go. The Turkish man led Baklava away to what I hoped was a home full of adoring children, like her brethren in Missouri, and Mel and I biked into another country. Goodbye, Turkey; *gamarjoba*, Georgia.

Other than the stamp in our passport, the only evidence we'd crossed some kind of border was the road: leading to the frontier it had been neatly paved, and leading away it was a cratered mudfest. When we heard a vehicle coming we scooted out of splashing range, but the driver swerved directly toward us. As he got closer I realized he looked oddly familiar. "*Why me why me*, hahaha!" the shopkeeper from Damal bellowed from an open window, steering his truck around puddles with one hand and with the other making that plucking-grape motion that seemed to signal approval. Mel and I waved back, then looked at each other and grinned: now all those doubting Turks in the tea house would learn we'd made it to Georgia.

6.

ANGLE OF INCIDENCE

Greater Caucasus

*F*lat light, flat mood. Flat road, too, except for the potholes, each of which cradled a puddle, which in turn cradled skies the colour and texture of porridge. At least it wasn't raining, only splashing up from below. Our wheels slurred along the wet gravel, leaving waves in our wake. I accelerated into murky puddles with no sense of what lay below the surface—a pothole, a bottomless abyss. Halfway across one I hit something and had to sacrifice a dry boot to stay upright. Icy water oozed between my toes and stole the feeling from my feet.

Georgia was torn between winter and spring, with a chill to the air that seemed on its way out. The landscape was the drab colour of decaying leaves. Some fields still had snow; others were partially thawed into dark bruises of grass and dirt. Every pasture was edged by thin, arching rows of trees. Large wooden crosses instead of minarets topped the hills—Orthodox Christianity swapping in for Islam. After Armenia, Georgia was the first state to convert to the religion, back in the fourth century BCE, and neither Mongol invasions nor Soviet imperialism have managed to shake the country's faith.

Eventually Mel and I came to a crossroads: we could bike on as planned to Tbilisi, the capital of Georgia, or go fifteen miles off-route to a small village where a friend of a friend was teaching English and had invited us to visit. Though we hadn't confirmed plans with him, we decided to make the detour.

A sense of purpose, however arbitrary, restores a dull road to its original lustre. Past the turnoff to Tbilisi we climbed steadily toward the village, the road paved now but still glittering with puddles. In a way those mirrors of water seemed more vivid than the landscape they reflected, as if the fact of a frame gave the sky and trees and hills a crispness and presence the actual world lacked. I pretended I could slip through to that more intense reality if I could only get the angle of incidence between my wheels and the water exactly right, like a space shuttle re-entering the Earth's atmosphere: too steep and you burn up; too shallow and you bounce back into outer space; somewhere in between, with the speed and tilt just right, you soar into another world.

Long before space flight, a pair of avid cyclists recognized the importance of the angle at which a wing meets the wind. Wilbur preferred long, languid rides on country roads while Orville loved racing, the faster the better. This blend of endurance and enthusiasm, and steadfastness and speed, enabled the Wright brothers from Ohio to soar where others had crashed, often fatally, including Otto

Lilienthal. The doomed "father of gliding" permanently fell from the sky in 1896, but Orville and Wilbur took inspiration from his achievements and decided to fashion their own flying machines using tools and parts from the bicycle repair shop they ran in Dayton. To make the wings they used unbleached "Pride of the West" muslin, a tight-woven cotton cloth ("fine as linen, soft as silk!") more commonly deployed in women's undergarments. For the ribs of the wings they used lightweight ash wood, and for the frame, lumber from a giant spruce. They tested the angles of incidence (also known as "angles of attack") of different wing shapes inside a homemade wind tunnel, and over years of trial and error realized—their eureka insight—that by warping wing tips in opposite directions, they could create different amounts of lift on each, making the plane tilt and turn. Previous glider designs forced pilots to steer by throwing around their body weight, meaning control was fickle. By belting the wing tips to their waist via wires and a harness, the Wright brothers could manoeuvre their Flyer by sashaying from side to side, the way you steer a bicycle at high speed less with handlebars than your hips.

For a launch pad the Wright brothers chose Kitty Hawk, a small town not far from where Mel and I finished our cross-country traverse on North Carolina's Outer Banks, where the soft landing of sand dunes and reliable winds offer ideal conditions for flight. Orville won the coin toss that decided which brother would attempt the first sustained, powered, heavier-than-air flight that cold December morning in 1903, when the wind blew in the Wrights' favour, which is to say strongly against them. Although Orville didn't fly far that morning—a hundred and twenty feet, or half the length of a modern Boeing 747 jet—the glory was in the details: a machine took off under human control, soared for twelve seconds under its own power, and landed as high as it launched, meaning it hadn't simply glided downhill. First flight.

That same day Wilbur bested his brother's pioneering flight by lasting nearly a minute in the air and travelling 852 feet, roughly the length of the Marco Polo Bridge in Beijing. Seven years after that, the world's first commercial cargo flight carried two bolts of silk from Dayton to Columbus, Ohio. And in 1969, roughly half a century later, Neil Armstrong brought fabric from the wing and wood from the propeller of the original Wright Flyer to the moon and back. I've always thought the moonwalker's pithy first words—"That's one small step for man, one giant leap for mankind"—more accurately described the Wrights' accomplishment. After all, Orville and Wilbur figured out how to fly from scratch, without institutional support of any kind, without a Mission Control monitoring their every move. I especially loved the fact that the brothers' success in flight had depended on a bicycle. The Wright Flyer, a homespun contraption of cloth and wood and wire, relied on a modified sprocket chain to power its twin propellers, and the hub of a bicycle wheel to launch off a railing and into history.

My own bike nearly launched off the road in Georgia, but only because the puddles turned to ice the higher we climbed. "Wish we had our riding boots!" Mel called out, a grin in her voice, and I was relieved she was in a good mood despite the dropping temperature. The two of us used to wear our scuffed leather riding boots to elementary school in the winter, not because the thin cowhide kept our toes warm, or because they were stylish by the already ruthless standards of the sixth grade, but because we'd discovered their treadless soles were essentially ice skates without blades. We'd taken turns towing each other across the frozen schoolyard at recess, seeing who could catch the biggest air on launch ramps built from snow. These were small acts of exposure, tiny flights into risk, and yet the world seemed a little larger each time we landed.

Mostly we crashed unglamorously into snowbanks, then dug ourselves out to do it all over again.

Cold toes were bearable then because we knew relief awaited at the end of recess, when we'd troop inside a warm school, and they were bearable now because we'd finish the day in a cozy Georgian home. But when we arrived in town and asked around for the friend of a friend, a gloomy, blunt-shouldered man claimed he wasn't there. He'd departed for a few days, or had never lived there to begin with, or possibly didn't exist at all—I couldn't quite tell which from the man's dismissive hand gestures.

Mel and I were wrong to stand there expectantly, waiting for an invitation to stay, but we'd been spoiled by hospitality in Turkey and dreaded the idea of backtracking. Icy slush gummed the cables and gears on the bikes, and when I wiggled my toes to warm them I only succeeded in sloshing frigid water around. The man stared at us without curiosity, as if into a great distance. When a grandmotherly woman shuffled over I was sure she'd be more sympathetic, but she just held her hand to her mouth in wordless surprise or perhaps dismay. Her age probably meant she had been a citizen of the USSR and had seen peaceful demonstrations for Georgian independence shot down by Soviet troops. She'd lived through civil wars and violent separatist movements, bread shortages and energy blackouts. She perhaps mistrusted the modern promise of neo-liberal democracy and wasn't wrong to do so; it was just the latest fad for organizing life in the South Caucasus, a subtle variation on the kingdoms, dynasties, and empires that had risen and fallen before. A gold ring hung on her thin finger, loosely orbiting the bone—a hand that had held hunger once and probably expected to grip it again.

Mel and I turned around and shivered back down the road we'd just biked up. At least it was downhill. Sometimes detours are the destination, and sometimes they're just detours. I didn't dry out from all the puddles until we reached Tbilisi days later.

Based on 1.8-million-year-old *Homo erectus* skulls found not far from the Georgian capital, the South Caucasus is one of the first places people (or hominids like us) migrated to out of Africa. The natural barriers of the Black and Caspian Seas, as well as the Greater Caucasus and Lesser Caucasus mountains, formed the indigenous borders of the region, parsing human communities more effectively than barbed wire and bureaucracy. In the relative isolation imposed by such rugged frontiers, different groups developed distinct quirks, preoccupations, and tongues, prompting Arabs to call the region Jabal Al-Alsun, or "mountain of languages." Even today the Caucasus Mountains (including the Russian side of the range) host one of the highest densities of distinct languages on the planet, with Georgians, Azeris, Armenians, Ossetians, Abkhaz, Kurds, Talysh, and Lezgins generally not speaking to each other in their mutually unintelligible tongues. Linguistic diversity is frequently coupled with biodiversity, and the South Caucasus is no exception. Some of the planet's richest and most threatened ecosystems are crammed into a chunk of land half the size of Manitoba. Somewhat ironically, the individuals dedicated to protecting these wildernesses are crammed into government buildings and NGO offices in Tbilisi.

Dressed in our very best synthetic hiking pants, Mel and I shuffled around the city interviewing these experts after applying for tourist visas for Azerbaijan, the next stop on our Silk Road. After a while one civil servant or government minister blended into the next, for when we asked questions about wilderness conservation they all answered with the same economic jargon about markets and incentives, or the same catch words of *ecotourism, natural resources,* and *sustainability.* But sustainable for who, what, why, and how long? Sustainable for the planet, or for the status quo of capitalism and consumerism? I had reason to suspect the latter, given that Georgia had recently tried to purge its Ministry of Environment by subsuming the Protected Areas Agency under the Ministry of Energy. The government eventually backed down, but only because it realized a dedicated Ministry of Environment would lure in more foreign aid money.

During these conversations I had the disconcerting sense that I was talking shop with Marco Polo. He and I both travelled to the same places by similarly slow modes of transportation, but the motives compelling each of us to and along the Silk Road were vastly different: Polo wanted to monetize and quantify whatever he could in the trading hubs along it, while I hoped to highlight the immeasurable worth of the places in between. To call us both "explorers" simply exposes the dangerous relativism of the term, its infinite co-optability, just as "wilderness conservation" was beginning to seem so malleable a concept as to be meaningless. When I asked about the latter I wanted to talk reverence, sufficiency, the economics of enough. I wanted to discuss the ways wilderness brings us alive in every possible sense. Isn't that the most obvious reason to take care of the planet, the fact that we can't breathe, drink, eat, or exist without it? Every attempt we've made to create a long-term closed ecological system—an artificial, self-sustaining planet-in-a-bottle in which all human waste products such as carbon dioxide, urine, and feces are sustainably metamorphosed into oxygen, water, and food—has failed, often expensively and dramatically, such as when Biosphere 2 in Arizona went hypoxic and suffered mass extinctions of plant and animal species, as well as population explosions of katydids, cockroaches, and ants. Knowing what we know now, we couldn't colonize Mars without serious and ongoing reliance on Biosphere 1. We need this world, and this world doesn't need us. Why do we persist in behaving as if the converse were true?

At least the NGOs we met with were a little more subversive in their thinking, a little more radical, particularly in how they worked across contentious borders in the South Caucasus. When formal nation-to-nation co-operation through a transboundary protected area or "peace park" proved impossible because of intractable conflict, civil society could sometimes pick up the slack by working through informal channels, below the political radar. In other words, scientists in nations that weren't exactly on speaking terms—such as Turkey and Armenia—could still discreetly share data and harmonize nature management strategies in a kind of "guerrilla" conservation effort. Such an approach has limitations, namely the restricted power of NGOs to formally change policy, but also certain advantages, given how nimble and adaptive civil society can be compared to sluggish post-Soviet bureaucracies.

Admittedly, I was also partial to NGOs because the Caucasus division of the World Wildlife Fund (WWF) let us camp in their conference room. Mel and I quickly decorated the businesslike room with our bike clothes and camping gear, draping sopping wet long underwear over chairs and unfurling our sleeping bags beneath the long central table. This set-up was somewhat reminiscent of my office camping days at MIT, though it offered proximity not to a lab but to nearby Georgian restaurants, bakeries, and cafés, which we frequented because the WWF-Caucasus headquarters lacked a kitchen.

No cuisine is better suited to the black-hole hunger of cyclists than Georgia's, which combines carbohydrates, dairy, and nuts in the most creative, delicious, calorie-dense permutations imaginable. One favourite was *khachapuri adjaruli*, a boat-shaped vessel of dough filled with molten *sulguni*, a briny traditional cheese, and topped with cracked raw eggs and hunks of butter. Another was *badridzhani nigvsit*, or eggplant and walnuts pulverized with lemon juice and garlic and spices in a rich pâté. We stumbled on these dishes by delicious accident or by admiring what someone else had ordered at an adjacent table, for the menus were written in Kartuli, the Georgian alphabet, whose thirty-three curlicue letters resemble the etchings of beetles in bark. English-translated versions weren't much easier to decipher. "Would you like spewed brain with mushpom?" offered Mel. "How about fried chick or minked meat with boiled bough, paired with battle wine?"

We ordered battle wine, obviously, as well as a host of other dishes with no English equivalent. The Romantic-era Russian poet Alexander Pushkin remarked, "Every Georgian dish is a poem," but it's also a song-and-dance routine, judging by how frequently poker-faced boys and men would get up after a restaurant meal to mince around on tiptoe wearing enormous furred hats. Our lessons in the challenges of translation continued back at the WWF headquarters, where we hung out with the security guards at night, watching television and drinking inky homemade wine. During commercials they tried to teach us some Georgian words, all of which sounded to me like choking on consonants, such as *gvbrdghvnit* ("you tear us into pieces") or *vprtskvni* ("I am peeling it"). Another favourite was *shemomedjamo*, meaning "I accidentally ate the whole thing," a word I'd needed on a few occasions involving baklava in Turkey. We also learned *zeg*, or "the day after tomorrow," which is what the Georgian secretary at the Azerbaijan embassy told Mel and me when we checked on our visa applications. I despaired a little at this, suspecting *zeg* was the Kartuli equivalent of *yok*—no, not yet, never—but amazingly, as promised, the visas were ready two days later and we hit the road.

Winter was melting away when we left Tbilisi in early March. Every tree branch was fisted with buds and the air smelled of freshly chopped pine, earthy and warm, all those rays of sunshine released. Every time I got on my bicycle after a long hiatus it was like riding back to myself, the only way there. The dissipation of life in the city—days of to-do lists, errands, emails, small talk with strangers— generated static in my mind that I didn't notice was there until I started pedalling and realized it was gone, the way you don't hear the hum of a refrigerator until it stops. Such is the paradoxical freedom of cycling the Silk Road. In restricting the range of directions you can travel, in charging ordinary movement with momentum, a bike trip offers that rarest, most elusive of things in our frenetic world: clarity of purpose. Your sole responsibility on Earth, as long as your legs last each day, is to breathe, pedal, breathe—and look around.

Rolling, forested foothills gave way to walled fields and wooden farmhouses. Some were painted a delicate robin's-egg blue, and their cracked walls gave the impression that something enormous inside was pecking through, ready to hatch. Practically every homestead boasted its own tangle of grapevines. Wine has been cultivated in Georgia longer than anywhere else in the world, by some estimates since the late Stone Age. Georgians seemed as self-sufficient in terms of growing their own food, for the shelves of shops were mostly bare except for an odd surplus of pickled vegetables and "Barf" laundry detergent. The villages were drab and dung-splattered but charged with life, especially when four young boys with flushed, ecstatic faces raced a horse-drawn buggy down one main drag, whooping joyfully as people leapt aside.

A few times in Turkey I'd secretly worried that I wouldn't discover anything new by slogging long distances on a bike again. What more could I learn by going through the same motions? Every day on a bike trip is like the one before—but it is also completely different, or perhaps you are different, woken up in new ways by the mile. If anything, the world grew more inscrutable the longer I looked at it, and the less focused I was on the brute mechanics of pedaling—aching legs and lungs, miles covered and miles to come—the more awake I could be to the world around me, its ordinary wonders. Like the bearded Georgian man I saw tossing straw with a pitchfork on a barn roof, presumably for insulation or storage, his grey hair and grey wool sweater fading into the cloudy grey sky so that he disappeared in the motion, became the stab and heave and fall of golden stalks himself. Or the way Mel would bike through a puddle and her tires would press wet fossils into the pavement, like the fine bones of Cambrian trilobites, only to evaporate and leave no trace. Or the face of the whiskery Georgian woman who sat by a wood stove in a small-town shop, her warm smile and watery blue eyes seeming to suggest that no road was long enough to learn all I wanted to know and get where I wanted to go.

Of course she was right, but I kept biking it anyway. We long our whole lives for things we've never known, places we've never been, abstractions that come alive to us in unexpected ways. What does the Silk Road have to do with Mars, except everything? Perhaps the great task of modern explorers is not to conquer but to connect, to reveal how any given thing leads to another: the red planet to the Silk Road, bicycles to the moon, a modern Georgian highway back in time to the Ujarma Fortress.

Once the fifth-century residence of King Vakhtang I Gorgasali, or Vakhtang the Wolf Head, who was said to tower above all others at a stately seven feet ten inches, the fortress is now in ruins that don't tower much higher. Mel and I wandered the maze of shattered rooms carpeted with grass and then decided to camp at the foot of the fortress—not because we were tired after only a few hours of biking, but because reading the afternoon away in the sunshine sounded like bliss. Mel was still marathoning through *War and Peace* on her e-reader, and I was turning the paper pages of Don Domanski's *All Our Wonder Unavenged*. This was the only physical book I'd brought on the Silk Road, and his words proved a perfect choice: they were inexhaustible, almost prophetic, as though Domanski were travelling the Silk Road a day ahead of me and writing about what he saw.

As the light faded we unfurled the tent on a rust of dry leaves. Among them was a pale, lemony-white flower that I guessed might be a Caucasian peony, or "Molly-the-Witch," named for the Polish explorer Ludwik Mlokosiewicz. He "discovered" this peony late in the nineteenth century, meaning he reported its existence to botanists in Russia and Europe (locals were of course well aware of the flower). But to his credit, Mlokosiewicz also recognized that this particular flower thrived on the slopes of the Greater Caucasus mountains and nowhere else. Realizing this was also the case for several other plants and animals, he presciently called for a nature reserve to protect these endemic species. A few years later, in 1912, the Russian Czar established what would become the Lagodekhi Protected Areas on the edge of modern Georgia, bordering Russia and Azerbaijan, a park we would visit in a few days. Meanwhile the flower grew precisely where we hoped to pitch the tent, so we carefully arranged the vestibule around it, in a kind of greenhouse, and shared a roof among the ruins.

~

Our next campsite wasn't nearly as bucolic, though it similarly rein-forced the fleeting nature of life when it turned into a firing range. Several Georgian men wearing camouflage showed up after we'd eaten dinner and shot at targets pinned to trees a few hundred feet from our tent. The hunters recoiled with the kickback of the gun—the testosterone-fuelled punch of it like guys socking each other in the shoulder in greeting—and we recoiled from them.

If life is chancy in Georgia today, though, it is tame compared to what *Homo erectus* faced in its venture out of Africa. As I slouched low in my sleeping bag I thought about the ancient hominid bones discovered in Georgia, with their flared cheekbones and chunky molars. Those early Pleistocene fossils were found cached in the former lair of a sabre-toothed cat, whose powerful canines fit like puzzle pieces into the holes puncturing the occipital dome of one skull. At MIT I'd attended a public lecture by a Harvard professor who referred to ancestral hominids as "those creatures" and mega-carnivores as "their faunal colleagues." The colossal scientific dis-tance in his words struck me as so absurd I jotted them down. It was as if he were holding up a human precursor between pincers for examination, though I rather liked the ring of "faunal colleagues" for the convivial atmosphere it implied in the ancient world, as though sabre-toothed cats strolled about in tweed and held office hours just across campus. Now all the megacarnivores have gone extinct, *Homo erectus* has evolved into us, and the fiercest threats humans face generally come from each other, a truth made pain-fully apparent as bullets decorated trees nearby. I hoped our own perforated skulls wouldn't be unearthed someday in the collapsed den of our tent.

As it grew darker the hunters departed and I could finally hear the river on whose bank we were camped. I lay there idly wondering what the Silk Road would be like in another million years, or even a thousand years, given the only firm fact is flux. Maybe the contested frontiers at Ani and Siachen would be mere historical footnotes, quirky details in some future student's thesis or a revised *Description of the World*, its maps and names almost alien to us. By what moniker would Georgia be known in the future? How would people refer to the border town of Lagodekhi?

"Lagoducky!" proposed Mel the next day, when a flotilla of ducks waddled at us like a welcoming committee. The eponymous national park was a steep climb out of town, and people cheered us on from the road shoulder as we biked there. A stone entrance gate and signs formally marked the Lagodekhi Protected Areas entrance, and just beyond it was a granite boulder showcasing a bronze portrait of Mlokosiewicz. We parked our bikes and spent the next two days hiking with a stout, root-like ranger named Giorgi.

Though he didn't speak English, Giorgi made the best of things by leading us to the nearest interpretive signs and pointing helpfully at the translations. He had salt-and-pepper eyebrows and a moustache to match, and dressed formally in a thick jacket, dinner shirt, and leather vest. There was a chill to the air when he led us into the woods, but Mel and I warmed up as we hiked, which meant we fussed continuously with layers to accommodate the temperature swings. Giorgi marched in his heavy attire regardless of sun or shade, movement or stasis, a steady-state weather system unto himself.

The forest was rayed with light, each sunbeam spliced by branches. Mossy trunks of oak and beech and maple towered over the trail. The interpretive signs disappeared, and I saw a man chopping wood and a woman collecting something from a hillside, filling a basket with whatever she found. I was confused: Wasn't this a park? But when I pointed the people out to Giorgi he looked unconcerned. Later that afternoon, a local Peace Corps volunteer would describe how Lagodekhi was divided into two zones, a small "Managed Nature Reserve" in the lower woods and a much larger "Strict Nature Reserve" in the alpine. The presence of humans in the park is mostly limited to the lower managed part, while the upper area is more like the *zapovednik* the entire reserve used to be under the Soviets. Back then the locals had collected wood and mushrooms from the strictly protected reserve anyway, the Peace Corps volunteer explained, and they grabbed as much as they could, to justify the stiff fine they'd pay if caught. With the managed buffer zone now permitting some legal collection from Lagodekhi, locals gathered only what they needed, knowing they could come back the next year for more.

The next day Giorgi took us to the Machis Tsikhe, a fifth-century summer retreat for bygone Georgian kings. The buildings were mossy and cracked with long, ragged lines. From inside the ruins, with daylight shining through the fissures, the walls looked scored with lightning frozen in its flash. Geologists define stones as rocks that have been put to use, so that, as the poet Don McKay put it, "What happens between a rock and stone is simply everything human." At Machis I found myself wondering what happens the other way around, when our careful curations of stone ruin themselves back to rock. Maybe we need a word for stone that has shirked its human duties. Or perhaps what happens between a stone and a rock is simply the consolation of not mattering much, only in the way my degree at Oxford didn't matter much, meaning it didn't deplete my time there of significance so much as free up its fullest possibilities: We're only here by fluke, and only for a little while, so why not run with life as far and wide as you can?

To the east, Giorgi pointed out the frontier with Azerbaijan, dim and suggestive, as though it hadn't mustered the force to wholly take form. The trunks of beech trees gleamed silver in the dark woods, and the dull light of a cloudy day gave everything a ghostly cast, particularly the two camouflaged men who materialized from the mist, lugging rifles. After a perfunctory glance at our passports, the Georgian border guards hiked along with us, happy for company on their lonely patrol.

Back at the park headquarters, we spent the evening attempting to interview another Giorgi, this one the head of administration at Lagodekhi Protected Areas. The dark-eyed, bearded man was a younger, swarthier version of our stalwart guide, but unlike ranger Giorgi, who didn't know any English, this Giorgi spoke "a leetle," which proved less communicative than none at all.

"So what kinds of endangered animals live in the Lagodekhi reserve?" Mel asked.

"No, no, no," said Giorgi dismissively. "There are no dangerous animals here."

"Sorry, I think we've confused you," I tried. "By 'endangered animals' we mean species at risk."

"No, no, you are not at risk, I am saying!" said Giorgi, indignant. "Ladies, there is *no dangers* in Lagodekhi!"

And on it went, for about an hour. At which point we gave up, thanked him, and walked away more bewildered than ever. I consoled myself with the fact that the Peace Corps volunteer we'd spoken to the day before had still seemed pretty lost despite living in Lagodekhi for nearly two years. "So what do people do here for a living?" I'd asked. "I'm not ... sure," he'd confessed with a helpless, homesick look.

In some ways it was the clearest answer we'd heard yet. It was certainly the most honest response he could've given. For other than exposing the obvious differences between a foreign land and wherever you're from—the way the Polish explorer noticed a peony growing in Georgia and nowhere else—travel reveals less about the truth of a place and hints more at how complicated the world is, how reeling and inscrutable. Perhaps that's the best thing going for bicycle travel in particular: the way it's an antidote to straight lines, haste. The danger of growing so sure of yourself that you forget you know nothing. Not the meaning of existence, not what's around the next bend, not even the way back to our room at the Lagodekhi park headquarters until ranger Giorgi showed us there.

Beyond the imposing Azerbaijani border gate, which we formally crossed the next day, a series of hulking, empty buildings had benches outside that were still wrapped in plastic, as though recently installed. Mel and I were hardly the first to cycle the Silk Road, nor would we accomplish anything noteworthy in an exploratory sense along its historic sprawl—but now I saw my chance! While waiting for Mel to take a bathroom break, I pulled away a strip of plastic on one of the benches and discreetly claimed the very first sit, because it was there. Then I rearranged the plastic and we biked on toward Zakatala, the *zapovednik,* or strictly protected area, that mirrors Lagodekhi across the Georgian border and also bumps up against Russia.

When there are no fences, no signs, it's hard to tell when you've arrived. A friend of the Peace Corps volunteer in Lagodekhi had given us directions, which verbatim consisted of "turn off at Mazix and ask for my friend Konul in Gobizara." We managed to find the turnoff at Mazix, which took us past stone walls and cottages that reminded me of the English Cotswolds, only they were grittier, more hardscrabble. Women in headscarves bent over the black earth, planting in soil freshly unburdened of snow. Soft-eyed cows chewed cud in the middle of the road, unperturbed by us biking within inches of their bony ribs. Nobody we spoke to showed even a glimmer of understanding when we asked, in a hopeful tone, "Konul? Gobizara? Zakatala?" *Niet*, they told us. *Niet, niet, niet.*

The gravel beneath our wheels grew into fist-sized rocks, then dissolved into mud. The road forked at one point, and when we stopped, a portly man wearing a Bauer jacket and blue jeans approached us. He had bangs but otherwise buzzed hair, and an enormous mole on his upper lip. We asked him our standard trio of questions and he listened with what seemed an unusual effort at understanding. Then he gave us the usual trio of answers. We opted to take the left fork because it looked as though it continued up the valley toward Russia, in what seemed the vague direction of the *zapovednik*.

Trees surged closer on all sides. The road turned to gravel and at one point vanished under a stream. The gravel turned to dirt, then an ooze of mud, then disappeared entirely where the woods thinned into a huge meadow. "In the middle of the forest," wrote the poet Tomas Tranströmer, "there's an unexpected clearing that can only be found by those who have gotten lost"—an apt description of our circumstances, except that this particular clearing was constellated with dung and close-cropped grass. It was more a cow pasture than a wilderness. We gave up on the *zapovednik*, pitched our tent, and refocused on food.

"Tonight's menu," Mel announced with a flourish, "features instant noodles, instant noodles, or, if you prefer, instant noodles."

"Perfect," I said. "What I've been craving all day."

As we boiled water on the camp stove, a pair of birds swooped above us, perhaps trying to glean whether Mel and I were dead enough to eat. Those dark, roaming knots of sky coasted toward Russian airspace, veered above us in Azerbaijan, and swooped toward Georgia, crossing frontiers with a few flaps of their wings. I thought of Halitherses, the elderly soothsayer in *The Odyssey* who boasted the unimprovable epithet of "keenest among the old at reading birdflight into accurate speech." Mel and I tried to guess at what the birds were scribbling on the sky. *Think beyond borders! Turn back while you still can!* Or maybe, *gvbrdghvnit,* "you tear us into pieces," which is what ecosystems say to fences—except at Ani, the Korean DMZ, and other frontiers that by chance let wildness flourish. If we read borders as narrative lines, sometimes they tell different stories than their authors intended. Sometimes the original plot runs wild.

We lingered outside after the meal, enjoying the quiet, the fact of thaw. A high ridge of the Greater Caucasus range, barely visible above the forest, burned coral in the fading light. Those mountains are synonymous with the myth of Prometheus, who stole fire from the gods and was chained to an icy peak as punishment—which didn't sound so punitive to me, except for the chain part, as well as the eagle eating his liver daily (since he was immortal, it grew back each night). Those minor details aside, who wouldn't want to live above the clouds, with stars tangled in one's hair like burrs? Then again, springtime down here had its perks, such as roads with reliable traction; mornings that didn't begin with blizzards, inside the tent or out; and less miserable travelling companions.

"Mel?"

"Yeah?"

"I'm sorry..."

For being secretly pleased when you found the going tough. For not forgiving you completely for high school. For eating more than my fair share of baklava on several occasions in Turkey.

"... sorry... that it was such a hard go for a while."

It wasn't a perfect apology, but Mel seemed to know it was heartfelt.

"The road can only go up from here. Like really up, to the Tibetan Plateau." She paused and fumbled around in the food bag. "Now how about dessert? On the menu is *this* tiny square of chocolate, or *this* tiny square of chocolate."

We savoured nubs of chocolate all the sweeter for their smallness as the sun sank behind the mountains, and when it was too dark to read birdflight into speech anymore, even the silence was like something winged.

7.

BORDERLANDIA

Caspian Sea

Zakatala the town proved easier to find than the nature reserve. The next morning we backtracked from the cow pasture after our failed mission to locate the *zapovednik* and rode our muddy bikes into a bustling town where everyone was dressed in black. I felt like a neon sign in my red synthetic jacket, and Mel was impossible to lose in her bright purple version and red curls. Without consulting each other we stopped when we saw a sign advertising baklava, or rather *пахлава*, as the pastry is rendered in Cyrillic. The sweets brought us right back to the Black Sea, rather unfortunately, for it was pouring when we left the restaurant.

Azerbaijan was reminiscent of Turkey not only in its weather but also in its abundance of tea salons. Lounging inside them was the usual profusion of jovial, unemployed men (never any women) who confidently informed us that Baku, the capital city, was 350, 500, or 230 miles away. Several of these tea-sipping Azeris asked to see our map—not to confirm the distance to Baku, I suspected later, but to see how their country, not much larger than New Brunswick, was demarcated on foreign maps, particularly the part of it known as the Nagorno-Karabakh Autonomous Oblast. This majority Armenian enclave voted to secede in 1990 from the Azerbaijan Soviet Socialist Republic, where arbitrary Stalinist borders had stranded it outside the equally arbitrary borders of the Armenian Soviet Socialist Republic, and the referendum led to a full-blown war. When the USSR collapsed under the weight of its centralized political and economic systems in 1991, fragmenting overnight into fifteen different nations, Nagorno-Karabakh—comprising roughly a seventh of Azerbaijan's total territory—was under de facto Armenian control and has remained so ever since. The ongoing conflict remains a sore spot for Azerbaijan, which regularly swaps crossfire with its nemesis neighbour in the form of bullets and also words: *AzerNews*, a popular daily English-language newspaper, features the usual sports, politics, and business sections, and also one dedicated to "Armenian Aggression."

I was relieved when our map provoked no outrage. We continued down the road to Baku, where Azerbaijan's former ties to the USSR are evident not just in the alphabet on signs but the country's apparent obsession with roadside concrete statues. Most of the ones we saw were of animals, and not species native to the Caucasus, but pink flamingoes, a tiger missing an ear, and a lion. We also saw larger-than-life statues of a man and woman saluting the USSR, or possibly God, or most likely the local equivalent,

Heydar Aliyev, "father of the nation," the first dictatorial president of Azerbaijan. He'd bequeathed the government to his son, Ilham, who loyally continues the family tradition of flagrant corruption and human rights abuses, though he hasn't yet gotten around to swapping in his own portrait for his father's on billboards. Every town featured a larger-than-life portrait of Heydar Aliyev beaming benevolently in front of an Azeri flag, which has the same eight-pointed star and crescent as the Turkish flag, only with a backdrop of blue and green stripes.

Fortunately the rain stopped a day later. In its absence the countryside seemed less drab than Georgia, the cows plumper, the grass greener. We followed a quiet, dreamy two-lane road that twisted and stretched like black taffy across Azerbaijan. The farther we pedalled east, the warmer it was when we woke up each morning, as though we were biking closer and closer to the sun. Every little town seemed more prosperous than the last, their streets lined with cellphone shops and Internet cafés, though the land beyond them quickly reverted to meadows and forests. Always to our left was the Greater Caucasus range, where mountains sleeved in ice held up the deep blue sky. Despite the region's conflicts ancient and ongoing, despite it having the longest history of human habitation outside of Africa, it was sometimes hard to tell where wilderness began and ended in Azerbaijan—with the exception, that is, of the country's designated roadside picnic areas.

Judging by how often these rest stops were packed with Azeri families, it seemed that picnicking in the sunshine was a popular pastime. With their cemented walkways, shin-high metal fences, overgrown lawns, and flowers wilting in concrete pots, the rest stops didn't exactly scream natural splendour, but at least they encouraged people to linger outdoors and appreciate the fresh air. Besides, the historian William Cronon argues that there is nothing "natural" about wilderness, that it is a deeply human construct, "the creation of very particular human cultures at very particular moments in human history." Though I might be appalled by Marco Polo's failure to swoon at mountains and deserts along the Silk Road, wilderness in his day implied all that was dark and devilish beyond the garden walls. The fact that I'm charmed by the shifting sands of the Taklamakan Desert and the breathtaking expanse of the Tibetan Plateau doesn't mean I'm more enlightened than Polo, more capable of wonder. It means I hail from a day and age—and a country and culture—so privileged, so assiduously comfortable, that risk and hardship hold rapturous appeal.

It probably also means I read too much Thoreau as a teenager. "In wildness is the preservation of the world," he wrote, priming me to pine after places as far away from Ballinafad as possible, like Tibet and Mars. Provoking such distant wanderlust was hardly Thoreau's fault or intention—he himself never travelled beyond North America—but I enthusiastically misread him, conflating wildness with wilderness, substituting a type of place for a state of mind. Cronon finds the whole concept of wilderness troubling for how, among other things, it applies almost exclusively to remote, unpopulated landscapes, fetishizing the exotic at the expense of the everyday, as though nature exists only where humans are not. This language sets up a potentially insidious dualism, for if people see themselves as distinct and separate from the natural world, they believe they risk nothing in destroying it. What Thoreau was really saying was that he'd travelled *wildly* in Concord, that you can travel wildly just about anywhere. The wildness of a place or experience isn't in the place or experience, necessarily, but in you—your capacity to see it, feel it. In that sense, biking the Silk Road is an exercise in calibration. Anyone can recognize wildness on the Tibetan Plateau; the challenge is perceiving it in a roadside picnic area in Azerbaijan.

By late afternoon these rest stops were generally empty. We pitched our tent in one on the way to Baku, but the lawn was so uneven we had to patch the holes in it with socks and underwear so the sleeping bags would lie flat. Using a concrete picnic table as a kitchen counter, Mel and I prepared our blandest meal yet, namely plain noodles mixed with flavourless flecks of corn from a soup mix. As I tried to pull the concrete bench closer to the table, for they were placed an awkward distance apart, I realized that these picnic areas, however neat and delimited, still hinted at the deep connectedness of all things: the bench wouldn't budge, being firmly anchored to the earth. "When we try to pick out anything by itself," observed John Muir, "we find it hitched to everything else in the universe."

The next day a young Azeri boy hitched a ride on my bicycle. He grabbed the rear rack and surfed along the road in his sneakers, refusing to let go even as I shouted and shook the bike. The boy eventually released his grip and ran off giggling, but I could've sworn he was still hanging on all the rest of that day, dragging his heels on steep climbs. The terrain flattened a few days later into a desert cut with canyons, sun-baked and glittering, a landscape that gave more literal meaning to the name *Azerbaijan*, which was derived from *azer*, Persian for "fire." It was so hot that I was sure the watermelon stand wobbling on the horizon must be a mirage. After stuffing our faces with fruit, the tailwind that picked up seemed no less miraculous. I was enjoying the boost of the breeze so much that I forgot the basic rule of flight: it's easiest when you counterintuitively go *against* the wind, not with it, which is perhaps why arriving in Baku felt like crashing.

Located on the western shore of the Caspian Sea, the oil-slick capital of Azerbaijan is the kind of place that sticks out its chest to hide its paunch. Until the early twentieth century, the oil fields of Baku supplied up to half the world's oil, padding the pockets of Azerbaijan's ruling elite. Because the city was so expensive we sought free accommodations through Couchsurfing.com, a website that connects travellers with hosts around the world. We accepted an offer to stay at an American student apartment that in photos looked vaguely bohemian, but in fact was just a dive. We unfurled our sleeping bags on the mouldy carpets of a room without a door, and between the students partying all night to the beats of Bob Marley, and getting snacked on by fleas or mosquitoes or both, we barely slept. Mel woke up with huge bags under her eyes and a swollen, rashy face, with bites freckling her freckles. I didn't need a mirror to know I looked the same.

We set off, bleary-eyed, to apply for visas for Kazakhstan, our next stop on the Silk Road, which was less a road, it turned out, than a long stretch of red tape. Forget traffic, weather, potholes, mountain passes: the hardest part about cycling from Turkey to India is getting permission to cycle from Turkey to India. We took a taxi to the Kazakh embassy in Baku, only to discover it had moved. We prowled around at great expense until we found the new location, but it was closed; a sign stated that the embassy's visa-granting hours were Tuesday through Friday, and it was Monday. Defeated, we caught another expensive taxi to our Couchsurfing apartment, but we'd neglected to write down the exact address. We got out in what vaguely seemed the right area, but we couldn't find the place after searching on foot for an hour, and our cellphone was dead, so we couldn't call our hosts for clarification. Having now spent most of our cash on fruitless taxi fares, we tried to withdraw money from half a dozen ATMs but they denied our bank cards. We tried to cash traveller's cheques at four different banks, but the tellers refused them. With just five manat left in our pockets (roughly equivalent to five dollars), we ducked into a dingy restaurant and ordered soup as an excuse to charge our cellphone.

Mel read over her logistical notes about applying for visas as we ate. In our division of responsibilities for the expedition, I was tasked with updating the website and fundraising, the latter of which wasn't going so well, given we were less than halfway through the trip and more than halfway through our money. Mel was responsible for visas and logistics, which so far had worked out just fine. But in the notoriously bureaucratic "Stans" of Central Asia, things were about to get trickier. Visas for Kazakhstan, Uzbekistan, Tajikistan, and Kyrgyzstan came with fixed entry and exit dates, meaning they had to line up perfectly in time and space, like moves in a chess game.

"Shit shit shit," Mel muttered.

"What? Is that a fly in your soup?"

"No! Oh goddammit yes." Mel fished out the fly with her spoon. "But there's an even bigger problem. So, uh, it takes longer than I thought to get Letters of Invitation for Uzbekistan." This was an expensive prerequisite for applying for tourist visas, one that allows Uzbekistan to screen its visitors and reap profits at the same time.

"Like how long?"

"So long . . . they won't be ready before our Azerbaijan visas expire."

I listened in stunned silence as Mel explained that extending tourist visas in Azerbaijan was complicated, and the penalty for overstaying a visa was severe. Even more worrying, the only Chinese embassy in all of Central Asia that was currently granting tourist visas was in Tashkent, the capital of Uzbekistan, making that country a crucial stop in terms of completing our planned route. We'd receive Nepalese visas at the border upon arrival, and we'd apply in Kathmandu for Indian visas, the final stamps we needed to finish this Silk Road—if we made it that far, which required sneaking across Tibet again.

I couldn't take my eyes off the flies twitching on the windowsill, reminding me of the Russian writer Isaac Babel's descriptions of flies in a jar of milky liquid in Tbilisi, "each dying in its own way." The same seemed true for how adventures perished in Baku.

"You're mad at me," said Mel.

"I'm not mad. I'm distressed."

"Right. At me."

"Not *at* you. *At* is such a direct word. I rarely feel things so directly."

I truly was distressed in all directions: at our lack of money to buy more than soup in this overpriced city, at the itchiness where bugs had savaged us in the night because we couldn't afford to stay anywhere decent, at the fact that we hadn't applied for Uzbek Letters of Invitation (LOIs) weeks earlier, in Lagodekhi, when I'd suggested we submit the applications and Mel reassured me there was no need to just yet.

Fortunately, urgent processing for LOIs was possible—for a price. With the cellphone now partially charged, we managed to reach the travel agent who was helping us obtain tourist visas in Central Asia, and he told us to meet his friend Elchin at 4:00 p.m. and hand him U.S. $160.

We taxied expensively around Baku in search of a bank that would cash traveller's cheques, finally found one, then drove at the appointed time to a sketchy part of the city where a skinny, dark-featured man in crisp blue jeans was slouched against a wall. He had the look of a man who buys high and sells low. "Are you Elchin?" we called out the car window. "Yessssss," he replied uncertainly. We handed him two crisp hundred-dollar bills and asked if he had change. He did not, but he "knew a guy." And with that he took off.

We waited half an hour before accepting we'd been duped. But just as we were about to leave we saw Elchin ambling down the sidewalk, in no particular hurry. He presented us with forty dollars in change and we thanked him effusively, grateful we could now afford to buy dinner. Little did we guess that this would be our life in Central Asia: a flurry of inscrutable transactions, enormous trust placed in strangers, sketchy arrangements to move a little farther down the Silk Road. We thanked Elchin once again and asked for a receipt. He just laughed, not unkindly, I think, but more as a hint that we were really pushing our luck.

"Something there is that doesn't love a wall," observed Robert Frost, yet equally something there is that does, for how else to explain their ubiquity, the unremitting brag of them everywhere? Whether buttressed with dirt roads or red tape, barbed wire or bribes, the various walls of the world have one aspect in common: they all posture as righteous and necessary parts of the landscape. That we live on a planet drawn and quartered is a fact most Canadians have the luxury of ignoring, for our passports open doors everywhere—with the notable exception of Central Asia, where North Americans face the kind of suspicion and resistance would-be tourists from Uzbekistan get from Canada, which offers a taste of life on the far side of the wall.

Kazakhstan's visa application process was fairly straightforward, though it involved a mountain of paperwork and patience. Kyrgyzstan's was similar—a little too similar, actually, for the embassy simply granted us another Kazakh visa, only with the country's name scribbled out and "KYRGYZ" stamped above it with ink that bled into other pages in our passports. But the fixed dates we were given the visas for didn't overlap, which meant our Uzbek and Tajik visas had to fill that gap or we'd be illegal refugees in Central Asia for a month. On a daily basis we called the travel agency, hoping for an update on the Uzbek LOIs, only to be told they were mysteriously "delayed."

Crossing the Caspian Sea was another hurdle. We could either fly or take a ferry to the far shore in Aktau, Kazakhstan. Boat travel was cheaper and more appealing, but we couldn't figure out when the ferry departed. Whomever we approached with questions at the docks delegated the matter to someone else, who would delegate it to someone else, and so on in a cascade of shirked responsibility. "Go, go, go! Always they say this, always what you want is further!" despaired Atlil, an Azeri-speaking Turkish friend of a friend who helped us navigate the shoulder-shrugging attitudes of bureaucrats in Baku. When we finally found the ferry *kassa* (operator), the surly Russian woman said she didn't know when the next ferry would arrive, when the last ferry had departed, or how long the crossings generally took, nor was it possible to reserve fares in advance. At this point even Atlil gave up. "Very problem," he muttered darkly, shrugging his own shoulders now, and it wasn't clear if he meant the *kassa* or us.

At least we were sleeping better. Mel and I had gratefully accepted an alternate Couchsurfing offer from an ebullient Mexican named Julio ("Rhymes with coolio!" he introduced himself) who shared a large, bug-free apartment with a Somali-Kenyan named Idris. Both were master's students at the Azerbaijan Diplomatic Academy, an institution whose very existence took me by surprise, given this country wasn't renowned for handling foreign affairs with tact and sensitivity, at least where Armenia was concerned. Perhaps my understanding of international relations was too by-the-book. A year after our visit, Azerbaijan made headlines for its "caviar diplomacy" with Council of Europe members, finessing their support by offering luxury vacations in Baku and lavish gifts of gold, silk carpets, and astronomically expensive fish roe from the Caspian Sea's depleted stocks of beluga sturgeon, whose "black caviar" sells for U.S. $2,500 per pound.

Since Julio and Idris's apartment didn't have Internet, Mel and I joined them sometimes at the Azerbaijan Diplomatic Academy, where we edited video footage from the trip into a three-minute teaser to post online in hopes of rallying enough donations to afford instant noodles along the rest of the Silk Road. In Baku, thanks to Mel's cooking, we ate everything but. "Hello, ladies!" Julio would cry whenever he came home with groceries, having usually called Mel three times from the store to confirm ingredients for huevos rancheros, Thai curry, pesto garlic bread, cheeseburgers and fries, and Chinese fried rice. Mel could work wonders when her kitchen wasn't limited to a tiny pot on a camp stove.

And so the days blurred by in Baku with no sign of Uzbek LOIs. The Caspian ferry came and went, so we booked flights to Kazakhstan for eleven p.m. on the day our Azerbaijan tourist visas would expire. We hoped this would leave enough time for Uzbekistan to grant us LOIs and visas, but there was still no sign of them the night before we were due to depart.

"If only we'd applied in Lagoducky," Mel said miserably.

"There's still tomorrow," I said with an optimism I didn't feel.

We searched the kitchen cupboard for the chocolate chip cookies Mel had baked the night before, but Julio confessed he'd eaten them all for breakfast. "Sorry, ladies!" he called out from the next room, where he and Idris were playing video games. So we made tea and drank it in silence, wondering where our Silk Road went from here. The night sky in Baku was pale and rheumy with light pollution, the city's steady bleed of photons into Cassiopeia and beyond, and as I stared out the window I couldn't remember the last time I'd seen stars.

The Uzbek LOI delay, it turned out, was entirely my fault. On the letter of employment part of the application I'd naively confessed to being a writer and even mentioned the freelance environmental policy work I'd done for an NGO after quitting MIT. I have no idea why I freely made these admissions, given Uzbekistan refuses to grant visas to journalists and expelled foreign non-profits from the country in the 1990s. Mel had prudently declared herself a student. As a result of my idiotic honesty, the Uzbek Ministry of Information was investigating my credentials and possible motives for travelling to their country, according to the travel agency handling our visa application process. Whether the agency had known the reason for the delay all along or had only just found out, they refused to say. Either way, the moral of this story is to lie through your teeth when it comes to borders and bureaucracy.

Mel had the enormous grace not to be mad at me, though my screw-up meant we barely scrambled out of Azerbaijan before our tourist visas expired. It didn't help that Azerbaijan Airlines made us unpack our bikes from the cardboard boxes we'd painstakingly procured and taped together in Baku, forcing us to expensively cocoon them instead in what looked like Saran Wrap. This thin pretence of protection meant the frames, gears, and wheels had no padding beyond the clothes we'd wrapped around them, but somehow the bikes landed in Aktau, Kazakhstan, in one piece.

The next day we vowed to continue bashing our heads against the nearest Uzbek embassy, now two thousand miles away in the former Kazakh capital of Almaty. "You're *sure* we can bring our bikes on the train?" Mel grilled the ticketing agent. Travelling with bicycles is far more challenging than travelling on them, so we'd debated leaving our cycling gear in Aktau because we planned on coming back here with our Uzbek visas anyway. But if we weren't granted visas, there was no point in returning, and leaving the bikes behind might tempt bureaucratic fate.

The travel agent reassured us that bikes were no problem. "*Da, da,*" she muttered.

"And there's food included with the ticket?" I asked. "Like meals served?" I assumed this would be the case on a seventy-two-hour sleeper train but wanted to make sure.

"*Da, da, da, da,*" she intoned wearily.

In these half-hearted reassurances we should've heard the ominous opening chords of Beethoven's Fifth. At the station the next day I guarded our bags on the platform in the pouring rain while Mel wheeled the bikes to the baggage car, where she was forced to bribe the attendant to get them on the train. For the second time in as many days we weren't sure if we'd ever see our bikes again, especially when he hand-scribbled a dubious receipt for Mel. Then we boarded only to learn, based on the vast quantities of food other passengers had brought, that there was no meal service. All we had was stale bread, a few apples, and the kind of peanut butter that lists peanuts last among its many ingredients, most of which I recognized from organic chemistry classes. But it was too late to bolster supplies. The doors clanked shut with amputative force and we creaked off to Almaty.

The train swayed drunkenly on its tracks and men swayed drunkenly through the cars. People were stacked in bunks like produce on shelves, some fresh, some overripe, most way past expiration. Outside, the desert was dull and wet with rain, a wide stare of dirt and grass. Blinking into view every few hours were concrete-block towns prowled by skinny dogs, all ribs and scuttle. I worried we'd look like them by the end of the ride, given our lack of food, although the air itself seemed caloric with vodka and fried dough vapours. Fortunately the Kazakhs across the aisle were as generous as they were prepared. Arrayed on the table in between their bunks was a feast of fried fish, boiled eggs, some kind of salty and oily soup, deep-fried bread, hard jewels of candy, and a skinned goat's head. They even brought porcelain plates and metal cutlery. After handing us forks and knives, they urged Mel and me to pare meat from the goat's brow.

Throughout the meal the Kazakhs chatted and joked with each other, their smiles so huge they absorbed eyes, joy triggering a brief blindness. A grizzled old man walked past me and playfully pinched my nose. Children swung through the aisles as though the train were a jungle gym, and I envied them their squirmy impropriety as we sat with the adults sipping tea. A red-faced toddler with curly blond hair and legs like stumps crawled into Mel's lap, crushing her quads. She had the pudgiest cheeks I'd ever seen. Later she wobbled down the aisle, kissing all the other little kids she could find, but the vast bumpers of those cheeks mostly prevented her lips from making contact.

We kept waiting for others to go to sleep, or at least the kids, not wanting toddlers to outlast us, but soon enough we gave up. Mel slid sideways into the upper bunk, which barely fit her from head to toe. "So cozy!" she giggled. I squeezed into the bunk beneath, its ceiling so low I could feel the heat of my breath bounce back onto my face. There were no curtains for the bunks, so we slept fully dressed. The sheets that had looked so crisp in plastic upon boarding the train were now clammy and wrinkled with our collective exhalations. I pulled one over me for a pretence of privacy.

Whenever the train stopped, the already hot air became

unbearably stagnant. I pressed my hands and face against the window, its glass luxuriously cold if damp with condensation. Rain pearled on the outside pane. The bunk was barely wider than my shoulders, and I had to brace myself to avoid pitching into the aisle as the train rocked along, making it hard to sleep. I thought about how Mel was forbidden from dozing in her parents' car as a kid, as were any visiting friends. "Look, kids," her dad would say, "you don't want to miss *this*," nodding at the same fields and forests we saw every day from the school bus. Mel still finds it hard to sleep in moving vehicles, and I wondered whether she was awake now. I didn't want to ask in case I woke her. Eventually I drifted off myself and dreamed about snakes, specifically about adopting a coral snake as a pet, which was strange because I wasn't fond of reptiles. My sheet slipped off while I was dreaming, but an older lady very sweetly draped it back over me. At least I assume that's what happened, because all I remember is waking up and kicking out in fright, convinced a snake was twining around my legs. The woman, startled by my reaction, beat a quick retreat down the aisle. "I'm so sorry," I called after her in the wrong language.

"You okay?" Mel asked, peering down at me from the top bunk. I explained what had happened and she nodded; it made as much sense as anything else on this trip, itself surreal and unpredictable from moment to moment. Was I already dreaming of snakes when the lady draped the sheet over my legs, or did the sensation of the sheet summon the dream in an instant? Does a dream anticipate or merely reflect a given reality? I often wondered the same thing about the Silk Road.

The next morning we sniffed out the dining car, a bright, breezy room with vases of fake red roses decorating all the tables. Mel and I ordered hot water to mix with Nescafé because we had scant money to spare for anything more substantial, and settled into the chairs with books. Before long a balding Kazakh man with beer on his breath joined us, despite the surplus of other available places to sit. He wore an off-white tank top with yellow armpit stains and stared dully into enormous distances, occasionally slurring a question at us in Russian while leaning in uncomfortably close. We pointedly ignored him until he went away and found his own table, where he ordered a large glass of beer. After a few sips he fell asleep.

Kazakhstan was a fast slide through sameness. I periodically turned the vase of fake roses so that their petals followed the sun. A woman walked by with an open briefcase full of shiny metal watches, but nobody in the dining car wanted to invest in watching time crawl. The slower the better, as far as I was concerned. The train ride offered a welcome suspension from stress and logistics, a blissful surrender to fatalism. Would the LOIs for Uzbekistan come through? Was the Silk Road as we'd dreamed it over? Would we ever see our bikes again? I sipped coffee that sloshed with the train and stared out the window as if I had nothing else to do in the world. Because, as I realized with relief and amazement, I didn't.

The steppe seemed to soldier on forever. Gaunt horses searched the dunes for spiky grasses with their soft lips. Once in a while industrial pipelines bridged the tracks in inverted U's. We passed rusting machinery and crumbling buildings, outposts that looked abandoned until little kids ran outside to wave at the train. We passed graveyards of fenced tombs crowned with crescent tin moons, the most solidly built structures around. And at some point, though I couldn't tell exactly when or where, we passed north of the Baikonur Cosmodrome, launch pad for the Russian space program.

I felt a twinge—not of regret, exactly, but of old ambitions shucked almost completely. I'd always expected to see this part of the world, but not from the fogged-over windows of a stinky, over-crowded sleeper train. In my childhood imaginings, I was on a Saturn V rocket launching for Mars.

Named for a town that is actually hundreds of miles away, just to throw off spies, this fenced-off spaceport was the launch pad for Sputnik, as well as the first man and woman in space—and, less gloriously, for an unknown number of intercontinental ballistic missiles. Astronauts rave about how they can't see any borders from low Earth orbit, yet the whole enterprise of space exploration is fuelled by a rabid nationalism. The same loyalty to arbitrary lines that sparked the Cold War also launched humans to the moon. How does cynical ambition, the capacity for mutually assured destruction, give rise to something as wondrous as a stroll on the Sea of Tranquility?

I thought of the Wright brothers, who shortly after making their giant leap at Kitty Hawk sold their plane to the highest bid-ding military, a fact I'd taken in at Oxford like a knife to the heart. But whether you blatantly sell out like the Wrights or not, all sci-ence and exploration carries a Promethean risk: if you steal fire from the gods, you can't predict or control how the flames will be put to use. The "survival of the fittest" mechanism behind Darwin's theory of evolution was adopted by the Nazis as eugenic justifica-tion for genocide. Fanny Bullock Workman's careful surveys of Siachen helped map the way to war over the glacier. When Galileo pointed what we now call a telescope at the rings of Saturn, the device was better known as a spyglass, and soldiers on Siachen use versions of them still.

So why keep stealing fire? When is enough knowledge enough? As I'd learned at Oxford, there is danger in viewing science and other forms of exploration as essentially noble enterprises. In that sense we're all positivists from the 1870s, convinced that with just a few more facts we'll figure it out, chart the ultimate map, engineer miracles to save us from ourselves. But "exactitude is not truth," as the painter Matisse put it, and the notion of science as a neutral search for it should not absolve scientists—or any explorers—of moral responsibility for the facts and maps they unleash on the world.

In the next booth the drunk Kazakh man who'd been quietly snoring suddenly woke up, chugged the rest of his beer, and passed out again. I stared out the window at a blue lake on the horizon, bluer by far than the sky. Closer up the wind-scuffed water looked like crushed velvet, and the lake was so huge the train seemed to slow down next to it, slogging along the shore at a pace I almost could've out-pedalled. Maybe the bicycle is the true pinnacle of all our science and longing to soar: one of the rare human inventions that has taken us farther, lifted us higher, without being warped to some sinister purpose. And for the moment I forgot about Baikonur as I wondered how my own bicycle was faring in the baggage car.

I found out when the train stopped in Almaty. The frame was scratched but intact, and Mel's bike was missing a handlebar grip. The baggage attendant, a different man than Mel had dealt with in Aktau, claimed he wasn't responsible for the damage because we didn't have a real baggage ticket. He laughed at the receipt Mel showed him. "This is what happens here, just accept it," said a Kazakh woman in business attire who was collecting her own damaged luggage. "Welcome to Almaty."

I was relieved we still had bikes at all. It was after midnight and raining as we awkwardly hauled them and all our gear over three sets of train tracks to the station building. I felt sore from sitting for days on end, subconsciously shifting my muscles to accommodate the train's movement. The taxis at the train station were too small for our bikes, so we commissioned the station wagon of a random family for the task. They drove us and our bikes to the address of Murat, a thirty-something Kazakh man we'd connected with through Couchsurfing. But the driver couldn't find the address, and we couldn't call Murat because we'd used up all our phone credit on the train pestering the travel agency about whether our LOIs had been issued (they hadn't). When we finally found the place, no one answered our knock on the door. Mel began chanting Murat's name, and the driver and I joined in, and eventually a tall, baby-faced guy shambled out, smiling through his sleepiness.

"My friends! Welcome!" he exulted warmly, though we'd never met before.

Over the next week Murat proved himself the patron saint of our trip. He spoke decent English and ran a travel agency with his parents, which meant he knew how to navigate all the spoken and unspoken rules of embassies in Almaty, which helped us score sixty-day tourist visas for Tajikistan. Sweet, slump-shouldered Murat wore a wide, serene smile and giggled at the slightest provocation. I could tell he hadn't understood something I was saying if he started laughing prematurely, before the period or punchline.

"And then this kind old woman on the train draped my sheet back over me—"

"Hehehe!" Murat chortled.

I didn't bother finishing the story. Instead I checked my email, and saw a message from Julio. "KATE, HEYA! SUP?" it read in all caps. "I SENT THIS LETTER TO MEL. I WANT TO MAKE SURE THAT SHE GETS IT. IN CASE SHE DOES NOT RECEIVE IT, COULD YA FORWARD IT TO HER? THANKS."

The actual letter read, "Dear Melissa, you are getting this email when you are in Kazakhstan on purpose, not out of cowardice but out of prudence: I did not want to make you feel uncomfortable and I did not want to feel ashamed." Julio went on to say that the way Mel made him feel "cannot be described with worlds [*sic*]." He realized that the odds were against him: ". . . you are in transit and have a boyfriend. I am unemployed, not sure where to go, and on the point of leaving the former Soviet Union." But regardless of how she felt in return, "I send you my love and blessings and wish you success in your travels so that you go back to your loved ones safe and sound."

Julio wasn't the only one along the Silk Road to fall for my inadvertent heartbreaker of a friend. Shortly after leaving Kars, Alkim had sent Mel a Facebook message confessing, "You stucked [*sic*] an arrow into my heart." This was accompanied by a link to a short video about our visit with KuzeyDoga, in which many of the shots lingered lovingly on Mel. Alkim was a filmmaker famous in northeastern Turkey, parts of Iran, and now small towns in Ontario where our parents lived, for the video had gone viral strictly among our families. Poor, arrow-stuck Alkim. Mel had written back graciously but firmly reiterating she had a boyfriend.

Mel was also checking email on Murat's computer, and when she exclaimed, "YES!" I was a bit startled by her reaction to Julio's confession. But it wasn't his love she was celebrating; it was an update from the travel agency: our LOI reference numbers had been issued.

We were frantic with joy until Murat scrutinized the documents more closely: the LOIs gave us permission to obtain Uzbek tourist visas at the embassy in *Baku*, not Almaty. Fortunately the travel agency reassured us that the issuing city could be adjusted. A few days later, after waiting outside in a line of cigarette-smoking Kazakhs, we handed our passports, application forms, and money to the Uzbekistan embassy staff. Twenty minutes after that, our Silk Road stretched wide open all the way to China.

If the visas worked, that is. Murat thought they all looked counterfeit. The Kyrgyz visa was a fake-looking stamp on a Kazakh visa. The Tajik visa was childishly filled out with blue pen. The Uzbek visa declared "THIS PERSON WILL NOT WORK IN UZBEKISTAN." Murat really cracked up over this last one. "Who would go to Uzbekistan *for work?*" he giggled, his huge sides heaving. I wished we had our Aliens' Travel Permits from Tibet to show him, purely to provoke that riotous laugh, so I told him about them instead. I only got as far as "and then we turned ourselves in to the police" when he roared prematurely—not because he hadn't understood, I suspected, but because surrendering oneself to the authorities in Central Asia was a darkly comic notion. I suddenly felt unnerved by what lay ahead, namely a country so suspicious of foreigners that we were mandated to register at hotels every night and had to surrender those receipts in order to leave the country. That was problematic, given we planned to camp across Uzbekistan. But we brushed off these concerns and boarded the Trans-Kazakhstan the next morning with Uzbek visas in our passports, the bikes formally checked into the baggage car, and enough snacks to last the trip, all thanks to Murat.

Crowding the aisles was the usual circus of chubby toddlers, families slurping goat's head soup, and boozy men dozing in the dining car. To the south the Tian Shan mountains towered from the horizon, and to the north the flatness was relieved only by the nodding heads of spring grasses. Eventually the mountains shrank away and the steppe turned buttery in the late afternoon light. The train stopped at a dusty station where plastic bags wafted like pollen on the breeze. People wandered the platform but seemed uninterested in boarding; perhaps they were just out for a stroll.

Women with shiny black hair pulled back in bright red and blue scarves sold dried fish from piles stacked on the platform, pyramids of scales that glinted gold and silver.

As the train clanked forward again a young man and woman with lofty cheekbones suddenly raced to get it, their faces lit up and full of laughter. Do people always grin like that, running to catch trains? A few minutes later they came through the dining car, holding hands and looking serious and ordinary again, and I wondered whether we're most alive in our moments of longing, the act of launching for a place we're not certain to land. As the train accelerated toward the Caspian Sea it made a clopping sound followed by a queer suspended pause, as if every fifth beat the wheels didn't touch the tracks, which somehow I hadn't noticed before. Out the window, barely six feet from the speeding train, a burro flicked its wiry tail at flies but otherwise didn't budge as we passed, and neither did the bird perched between the burro's ears.

PART THREE

∿

I would like to do whatever it is that
presses the essence from the hour.
ELLEN MELOY,
THE ANTHROPOLOGY OF TURQUOISE

8.

WILDERNESS/WASTELAND

Ustyurt Plateau and Aral Sea Basin

G etting off the train in Kazakhstan felt like starting the Silk Road afresh, only properly this time, because we didn't miss our stop. People mobbed the platform in Beyneu, little more than a stretch of bare dirt bordering the rails, which vanished in a hot blur in both directions. We unloaded our gear and bikes in the only empty space we could find, against a whitewashed building flanked by poplars, then Mel went off in search of water. Only when the wind changed direction did I realize why such an enticing, shady spot had been vacant, and why those trees were flourishing in a desert: we'd parked ourselves next to the public restroom.

The stench didn't stop an ample-waisted Kazakh woman from

shuffling over to me. The red velour bathrobe she wore gave her a dishevelled but oddly glamorous look, like an opera singer who'd just woken up. The woman gestured at my bike and made pedalling motions with her hands. I shrugged and smiled, offering her the bike. Below her head scarf a huge grin swallowed her eyes, two dimples in rising dough, then she sped off with alarming speed. Men in pit-stained tank tops and dusty suits stepped aside to let the bike pass, then the crowds closed behind her.

"*Velocipede*, BYE BYE!" someone said, laughing.

My bike was vagabond, Mel was who-knows-where with our water bottles, and the toilet fumes could've knocked the train off its tracks, so why did I feel like singing an aria? Because just beyond Beyneu was the Ustyurt Plateau, a desert stretching from Kazakhstan into Uzbekistan, and after a month in the bureaucratic doldrums, including 144 hours on a train, we finally had permission to go there. I was prepared to walk the sixty waterless miles to the Uzbek border, if that's what it took. Fortunately the lady in red returned my bicycle, Mel struggled back with several camel humps' worth of water in aptly named dromedary bags, and we set off spinning into the desert's immaculate yawn.

Even in April the heat was scorching. With no buildings or trees or clouds to block the sun, it blazed down on us with undiluted intensity. The road was less a feature of the desert than its erasure, a bare strip of land in a land itself mostly bare, except for a six o'clock shadow of grasses and herbs. Salt frosted the dirt in a cruel illusion of coolness, the result of brackish water evaporating in the millennia since the Ustyurt Plateau lay at the bottom of the Paratethys Sea, a breakaway remnant of the Tethys Ocean. The Paratethys drained away as the Tethys slipped under the Eurasian continent, uplifting Central Asia in the process. Now the sloshing of the drom bag on my bike was the only echo of the ancient shorelines I could just make out on the horizon, where the putty-coloured desert flared up into ridges of clay and gypsum.

The dirt road had been concussed into concrete-hard peaks and valleys by transport trucks hauling goods to Uzbekistan. Fortunately traffic was scarce, and we could avoid it entirely on the alternate tracks that snaked beside the main road, where vehicles had veered off in a fruitless search for smoother driving. As I cut across the desert to one of these sidetracks, I could smell the sage crushed beneath my wheels, a fragrance I blissfully associated with sneaking out of the Mars Desert Research Station. I stopped and fixed a bouquet of the herb to my handlebars, so that winds all across Uzbekistan would waft the spice of freedom into my face.

"This reminds me of Utah!" I exulted to Mel, who was biking next to me in the parallel tire track. It was rare that we could ride side-by-side without fear of being run over.

"Right, Utah," said Mel, suddenly very serious. "Where you went to space camp that one time . . ."

"It wasn't space camp!" I protested.

"Ah, my mistake, *the Mars simulation.*"

Merciless teasing was Mel's deepest expression of fondness. We were having a grand time, as if the widened horizons of the desert were not just a fact of geography but a mood. The Ustyurt Plateau reminded me of Utah, Ladakh, the Gobi Desert, the Taklamakan Desert, the Tibetan Plateau—basically every place that had ever mugged me with its beauty, its sense of peeled-wide possibility. When a reporter asked Orville Wright to describe where he'd flown the world's first airplane, he dreamily mused that the Outer Banks was like the Sahara, or what he imagined the Sahara to be. Travel is perhaps one part geography, nine parts imagination. You launch from the Outer Banks and land in the shifting dunes of northern Africa. You set off for Mars and end up—marvellous error!—on the Silk Road, this conjuring of dust and light and desire between Europe and Asia.

Marco Polo never knew this trade route by its modern name. "Silk Road" was coined by a German geographer in 1877 to describe the flow of goods and ideas between East and West. The term remained fairly obscure until explorers in the mid-1900s recognized its romantic appeal and slapped it onto book covers about travels in China and Central Asia. Today the Silk Road remains a clever marketing ploy, a catchy name to connect the dots of a holiday and lend it a kind of historic momentum, for tourism is one part geography, nine parts souvenirs and selfies in front of ancient monuments.

There were no such monuments on the Ustyurt Plateau, just the desert itself, which pleased Mel and me enormously. Much fuss is made of the distinction between tourists and travellers, particularly by those who insist, perhaps a little too strongly, that they fall in the latter, supposedly less superficial, category of foreign experience. Mel and I weren't camera-toting consumers of packaged sights; we were seekers of the raw and the real! At least until we reached the Kazakh-Uzbek frontier at dusk, where a perspiring Uzbek border guard frowned at our passports and questioned our motives for visiting Uzbekistan.

"Journalists? NGO? Correspondents?" he suggested with a sly look.

"Tourists!" we chorused brightly.

A Cyrillic sign welcomed us to OZBEKISTAN. This nation's similarities to the land of Oz would prove more than etymological, starting with the road beyond the border, which was so gilded in the setting sun it looked paved in yellow bricks. That it was paved at all seemed a feat of wizardry. Hiding behind the iron curtain, in Uzbekistan's case, was Islam Karimov, an undemocratically elected despot with a flair for corruption, torture, and forced child labour in state-owned cotton fields. Though he has passed away since our trip, Karimov at the time commanded such fearful deference from his citizens that when his cavalcade was due to drive past some cotton fields that had already been harvested, the locals glued wads of cotton back onto the plants so the president could behold the nation's bounty. Such rancid paranoia suggested that Mel and I should take seriously Karimov's decree that foreigners register with the Office of Visas and Registration (OVIR) within seventy-two hours of entering Uzbekistan, and judiciously collect receipts from hotels every night, without which we wouldn't be allowed to exit the country. At the very least, it seemed prudent to conceal our campsites. But in the desert there's no such thing as out of sight.

We pitched the tent a mile or so from the road on dirt cracked into neat polygons, as if by some enormous impact, possibly the bricks of Uzbek currency we dropped there. The single U.S. hundred-dollar bill we'd handed to black-market traders just inside the border had yielded hundreds of bills of Uzbek som, so many that they were held together with elastic bands, just like in the movies. It would be nine days before we found a hotel willing to relieve us of some of this money, in the form of bribes for doctored OVIR receipts that testified to us staying in sanctioned tourist accommodations. For now we buried the relative fortune deep in our panniers and tried to ignore the extra weight.

The next morning we inched across the Ustyurt Plateau as though *velocipede* were not just the Russian word for bicycle but a historically accurate description of the contraptions we rode. Anyone observing Mel's and my slow pace would've readily believed our bikes lacked pedals, tires, and drive trains like the cumbersome two-wheelers fashionable on the streets of Paris in 1876. Our hard-earned leg muscles and lung capacity had atrophied during our month-long hiatus from the bikes, meaning the previous day's jaunt of sixty rough miles from Beyneu to the Kazakh-Uzbek border was akin to running a marathon after weeks of bed rest. Even worse, the border road's washboard texture had substituted bruises for where our butts used to be, so that Mel and I howled in unison whenever we mounted our bicycles. We resorted to pedalling standing up, which made for slow progress. I consoled myself with an old Arabic saying I'd read somewhere: the soul invariably travels at the speed of a camel.

By that measure we were right on pace. At one point we passed a pair of camels with flopped-over humps and shaggy dreadlocked hides, nibbling with pursed lips at what looked like pure dust. Judging by the letters *AL* spray-painted on their ribs, as well as the jaunty scarves tied around their necks, these camels were domesticated, being one of the few large species that can survive the blistering heat and aridity of the Ustyurt. Another is the saiga antelope, an ungulate with dual unicorn-like horns that I thought I kept spotting on the horizon, only to see the four-legged creatures disappointingly resolve into more camels. This wasn't unexpected, for saiga populations have been decimated by illegal poaching for their horns, which are prized as aphrodisiacs by traditional Chinese medicine, the form of "healing" most harmful to life on Earth. The herds also periodically suffer mysterious mass die-offs, possibly due to bacterial infections from domestic animals. As a result, saiga are rarer than free speech in Central Asia. Based on photos I'd seen, their bulbous brown eyes and nose like an abbreviated elephant's trunk made them look Dr. Seussian, even extraterrestrial. But who was I to judge? From the perspective of Uzbeks, Mel and I were the aliens in the Ustyurt, a pair of pale, mute barbarians who opted to bike for nine days across a desert that locals sped through by car or avoided entirely. Then again, the Uzbek language has no word for "fun."

Before long our bottles and drom bags were empty. With no buildings or towns in sight, we resorted to flagging down vehicles for a resupply. The sky spilled onto the pavement where the road met the horizon, giving distant cars the appearance of driving on air. Regular cracks in the road made them sound like rapid-fire machine guns, an accelerating *tat-tat-tat-tat* that made us want to duck for cover. Instead, Mel and I stood taller and waved our arms. One tiny sedan was stuffed with peach-skinned mannequins, so that dozens of armless, legless, and headless torsos sprawled on the roof and crawled from the trunk. Mel and I dropped our arms and let it pass. Later came a Kazakh transport truck, and though the driver didn't have any water, he generously gave us a dented bottle of warm Coca-Cola. I went to twist off the cap but the seal was already broken. We guzzled half on the spot and saved the rest for dinner. "I wonder how instant noodles will taste," Mel mused, "when boiled in Coke." Fortunately we didn't have to find out. A few hours later we caught up with a crew of road workers who seemed delighted to fill our water bottles and bags—any excuse to stop shovelling tar in the smouldering heat. We posed for photographs with them and they thanked us effusively, as if we'd just given *them* water in a desert.

To avoid the worst of the heat, as well as the headwinds that did nothing to relieve it, we fell into a rhythm of pitching camp in the early afternoon and waiting until the world cooled and stilled before pedalling off again. Even with the tent doors opened wide to the wind, and sleeping bags draped over the roof to generate shade, the Glow-worm was an insufferable greenhouse. Veins popped out on my skin, charting hot rivers of blood. My hands and feet in particular felt oddly thick with heat, as if pressure were building up beneath my palms and soles with no outlet for release. It was too hot to sleep, to talk, to write, to exist. I lay on my sleeping pad and read as sweat pooled under my shoulders.

Among the books on my e-reader was a collection of John Berger's essays, but maybe the heat corrupted the file, for throughout Berger's text the word *the* was systematically replaced with *die*, as in: "Die third dimension, die solidity of die chair, die body, die tree, is at least as far as our senses are concerned, die very proof of our existence. It constitutes die difference between die word and die world."

But if the word and the world only differ by the letter *l*, an even slimmer border separates a desert from desertification in Uzbekistan. Mel and I were steeping in wildness on the Ustyurt Plateau, darkening in it like tea, but a few hundred miles east was a similarly parched-looking landscape that used to be the Aral Sea. Once the fourth largest lake on the planet, it was now a social and ecological wasteland. Historically fed by two rivers, the Amu Darya and Syr Darya, the lake had withered away because of industrial-scale irrigation in Uzbekistan, which began in the 1960s as a Soviet project to cultivate cotton, one of the world's thirstiest crops, in a desert. Thirty years later, the rivers no longer reach the lake, which has lost 90 per cent of its original volume and quadrupled in salinity, resulting in the catastrophic collapse of almost thirty fish species and, with them, the Uzbek fishing communities that used to harvest them.

"You cannot fill the Aral with tears," wrote the exiled Uzbek poet Muhammad Salih, and technically he's right: they aren't salty enough by current salinity measures. If you were unaware of the recent history of the Aral Sea, though, you might get duped into believing the stark, resilient ecosystem of the Ustyurt simply

extends eastward to the former sea floor, a sweep of sand incremen-
tally lighter in colour than the surrounding terrain, like scar tissue.
Locals call that new desert the Akkumy, meaning "white sands,"
and apparently only a graveyard of wooden ships stranded among
the dunes betrays the landscape as freakish, unnatural—not a genuine, living desert but the result of catastrophic desertification.
Maybe that's the true distinction between a wilderness and a wasteland: the latter is of our own making, rendering the Earth a little
more sterile, a little more arid, a little more like Mars every day.

I thought I spotted the red planet when we set off biking after dusk:
a bright ember on the horizon like the lit end of a cigarette. Mars—
if it was Mars—looked so tiny I could redact that entire world with
my pinky finger, the way Neil Armstrong said he could blot out
the Earth from the moon with his thumb. "Did that make you feel
really big?" someone asked him upon his return. "No," the first
moonwalker confessed in a rare candid moment. "It made me feel
really, really small."

Uzbekistan had a similar effect on me, especially after the sun
went down. In that world suddenly cooled no speed seemed
impossible, no destination too far-fetched. The desert was so flat
the constellations came right down to Earth, and stars hovered all
around me at eye-level, so it seemed like I was travelling to them,
even through them. With every pedal stroke I was soaring closer to
Saturn's rings, I was surging past the heliopause, I was flying neck
and neck with the Voyager spacecrafts as they sped toward Sirius,
a sun just 8.6 light-years from our own.

Tellingly, I could never muster as much interest in the Voyagers'
scientific mission as in their more whimsical cargo: mounted on
the instrument bay of each probe is a Golden Record, a twelve-inch
gold-plated copper phonograph inscribed with earthly greetings in

fifty-five tongues, snapshots from around the world, and a medley of natural sounds and human music: whale song; waves breaking on a shore; the heartbeats of a woman in love; pictures of Oxford's spires and towers, and of a modern airplane taking flight. The biologist Lewis Thomas suggested including the complete works of Johann Sebastian Bach, "but that," he admitted, "would be boasting." So they settled on just three of Bach's compositions, as well as the first two bars of Beethoven's Cavatina, supposedly the only music that moved the deaf composer to tears. Also etched onto the record are recordings as diverse as Azeri bagpipes, a Navajo night chant, and a Morse code rendering of the Latin phrase *ad astra per aspera*, "through hardship to the stars." Such fragments present— to any extraterrestrial intelligence that might discover and decipher them—a partial survey, a kind of haiku summary, of the unguessable gamut of life on Earth.

The Golden Record, as far as I can recall, wasn't discussed in the grade school science class that introduced me to the Voyager spacecraft. The discs weren't science, after all, but something else, more like poetry, though I don't think they came up in my English classes either. Instead I learned about the record early in university, from a book by Carl Sagan called *Murmurs of Earth: The Voyager Interstellar Record*, which gave me shivers with its hope that, "like Marco Polo, [the Golden Record] will find itself at the gates of some ancient and great civilization." At that point I was still smitten with Polo's Silk Road explorations—meaning still blissfully ignorant of their mercantile leanings. But if Polo fell in my esteem, Sagan remains an idol, an explorer equal parts scientist and poet. He led the team responsible for compiling the Golden Record, charged with making its selections fair in terms of geographic, ethnic, and cultural representation while giving a comprehensive overview of life on Earth. In the process Sagan slipped in some revealing humour. A picture from Sir Vivien Fuchs's 1958 trans-Antarctic expedition shows a Sno-Cat—the love child of a monster truck and an army tank, which the team attempted to drive across Antarctica—tottering precariously over the edge of a massive crevasse. "Freeing stuck vehicles may be an experience we share with alien explorers, no matter how advanced," noted Sagan. He also included a photograph of someone climbing a jagged spire in the Alps. "If the recipients recognize the silhouetted human figure, they may guess that it was both difficult and seemingly pointless to scale this rock needle. The only point would be the accomplishment of doing it. If this message is communicated, it will tell extra-terrestrials something very important about us." Possibly that humans are not entirely rational, though we often try to pretend otherwise.

The Golden Record itself was a seemingly irrational project: Why waste valuable payload space on a time capsule when it could be otherwise devoted to more fruitful scientific instruments? NASA even hesitated to take the "pale blue dot" photograph, for it required turning the spacecraft around (which was costly in terms of energy) and pointing a camera at the sun (which risked frying its optics). Such a photograph wasn't sensible, in other words. It was a frivolous use of taxpayer dollars. It wouldn't reveal anything novel or groundbreaking, given we already knew without a doubt where our home planet was situated in the solar system and what it looked like.

Science, sadly, has never been overly concerned with self-reflection. Only passionate lobbying by Carl Sagan managed to change NASA's mind about the photograph. What those administrators and engineers didn't seem to realize is that exploration is a systematic inquiry into the nature of things but also a radical, revealing art, much like science itself. Though exploration might result in new territory conquered or a heightened mastery over the material world, its real value lies in how it expands our consciousness, our sense of connection with each other and the universe of which we're a part. Which means that seemingly impractical gestures such as the "pale blue dot" photograph or the Golden Record itself are not diversions from serious exploration but its *essence*. What is the point of exploring if not to reveal our place in the wild scheme of things, or to send a vision of who we are into the universe? A self-portrait and a message in a bottle: maybe no other maps matter.

During those night rides across Uzbekistan I felt like a message in a bottle myself. I tried to make it out, but all I could read was the thin scroll of road lit by my headlamp. Something about potholes, about grass shooting up in the pavement cracks. Something about how there are places you can get to by road, and places you can only get to by *being on the road*—a state of mind you can bring to almost any context, especially a highway paved in stars in Uzbekistan. At least until you're so sleepy you nearly fall off your bike.

At about two in the morning Mel and I would pitch the tent by headlamp, doze for a few dreamless hours, then wake before dawn and launch into outer space all over again, pushing hard to make progress before the sun rose and the moon set. That first week it hung in the sky like a half-eaten rind of fruit. *What was it like to walk there?* people asked Neil Armstrong all the time, at the grocery store and the barber shop, their faces lit up, expectant. *What was it like to be first?* I would've done the same had I ever met him, knowing even as I voiced such questions that they were profoundly unoriginal, but wouldn't any other line of small talk seem frivolous? I'd ask in the ludicrous hope that Armstrong, having gone where no one else had been, might offer me something no one else could: a different map, a new religion, souvenirs the colour of stars turned inside out.

In any case, the famously shy moonwalker didn't answer. Maybe Armstrong's silence on the subject was a courtesy, a kind of imaginative generosity, as if this deeply modest man sensed that anything he could say would fall short of whatever people could dream. Or maybe he was so busy on the moon he didn't have time to absorb the marvel of being there, with his every heartbeat micromanaged by NASA. The checklist sewn onto the cuff of his spacesuit glove called for Armstrong to take photos, inspect the condition of the "Eagle" Lunar Module, hammer and scoop and bag samples of lunar rocks and dust—basically do everything but ponder his surreal place in the cosmos. Plus "the right stuff" in NASA's early days included the terse courage of test pilots, but not necessarily a capacity for evocative expression, which in retrospect seems a missed opportunity. After all, the Latin root of the word explorer is *ex-plorare*, with *ex* meaning "go out" and *plorare* meaning "to utter a cry." Venturing into the unknown, in other words, is only half the job. The other half, and maybe the most crucial half for exploration to matter beyond the narrow margins of the self, is coming home to share the tale.

In high school I'd been obsessed with the science-fiction movie *Contact*, adapted from the eponymous novel by Carl Sagan. The main character in the film is Dr. Ellie Arroway, an edgy, whip-smart astronomer who is obsessed with searching for signs of intelligent life in the universe. "If it is just us," she reasons, "it seems like an awful waste of space." One day in the New Mexican desert, where the Very Large Array of radio telescopes listens patiently to the cosmos, Ellie picks up a faint but unmistakable transmission from the vicinity of the star Vega, twenty-five lightyears away. The radio signal is eventually decoded into a blueprint for a machine with a pod for human transport, though nobody knows where the pod will go once the machine is activated. After a gruelling selection process, Ellie is chosen to find out.

Upon ignition, the pod plummets into a wormhole—a tunnel through space and time as predicted by general relativity—and takes Ellie on a warp-speed tour of the cosmos, past black holes, spiral galaxies, and other living, breathing worlds. During this journey, and despite her rigorous scientific training, Ellie doesn't reach for some kind of measuring stick or a compass or other instrument with which to quantify the experience, nail it down in data. What she reaches for is words. If only she can get them right, she knows life on Earth will be transformed forever, for what she sees and desperately wants to communicate is that the universe is more vast, resplendent, strange, and alive than we can possibly imagine, that we belong to something far greater than ourselves, that none of us is ever alone. She longs more than anything to share that awe, that humility, that hope. So she reaches for words, but like Armstrong she can't find them. "They should've sent a poet," she whispers instead.

When delegates from the UN Outer Space Committee recorded greetings for the Golden Record, many included lines from their countries' poets or spoke poetically themselves, such as the Nigerian who described his home continent as "a land mass more or less in the shape of a question mark in the centre of our planet." I often thought of his words as I hunched over my handlebars on those dark desert rides, feeling like a question mark myself in the centre of the universe, because wherever we go, there we are, even on a bicycle slogging slowly across Uzbekistan.

Deserts have long been landscapes of revelation, as though the clean-bitten clarity of so much space heightens receptivity to frequencies otherwise missed in the white noise of normal life. This was especially true just before dawn on the Ustyurt Plateau, when the horizon glowed and shimmered like something about to happen. As the sun rose it tugged gold out of the ground and tossed it everywhere, letting the land's innate wealth loose from a disguise of dust. The air smelled of baked dirt spiced with dew and sage. Our bicycles cast long cool shadows that grew and shrank with the desert's rise and fall, its contours so subtle we needed those shadows to see them. The severity of the land, the softness of the light— where opposites meet is magic.

In those cool hours before dawn the chinstrap of my helmet was stiffened with salt, a rough blade at my throat until I sweated enough that it softened. It didn't take long. As the sun climbed higher it seemed to roar, but that was just the wind rising with it, building in heat and intensity until biking felt like re-entering the Earth's atmosphere at a viciously steep angle. The horizon went from lovely to lurid, a red smear of lipstick on the rim of sky. By nine a.m. all the beauty and benevolence of the desert burned away, leaving a landscape seared of nuance and detail, and our pace became so slow that Mel and I risked being confused for roadkill.

Surely that's what the steppe buzzard had in mind when it circled above us for hours, barely flicking its broad wings yet soaring, mockingly, as if the whole sky were a downhill slope. Far below, spinning wheels across Uzbekistan's formal, blazing flatness, I smelled the real casualties of the road long before I saw them. A hint of rot on the breeze was followed by a flattened lizard, or a smashed-up hedgehog, or a tortoise whose cracked half-dome looked like a puzzle of the Earth with a few pieces missing. The wreckage was nothing compared to what I'd seen biking across America. I was somewhere east of Carson City in Nevada when the road began to shine. At first I thought it was the sun setting behind me, spilling slick light all over the tarmac. Then the noise started. A snapping and crunching like popcorn under my wheels, only the kernels were brown and endowed with six or fewer frantic legs. Locusts. Thousands of them.

Technically, I learned later, they were Mormon crickets, a well-armoured breed of katydid whose numbers explode in creepy, cannibalistic swarms following a drought, which is when I happened to pedal across the Silver State. The highway was a hard sheen of chitin roamed by a horde of half-squashed monstrosities, some missing legs, others wings, their rigid exoskeletons deformed by the hot press of wheels. That night my pasta dinner went cold as I tried and failed to muster an appetite. I had difficulty locating a patch of ground uninfested enough to set up camp, and fell asleep to the nightmarish patter of tiny legs on the tent roof. Insects, many of which are adept and fuel-efficient flyers themselves, inspire significantly less envy than birds.

What they have inspired, oddly enough, is travel literature, at least in the case of Wilfred Thesiger. The British writer and explorer penned his masterpiece *Arabian Sands* after working for the British Middle East anti-locust unit in Arabia's Rub al Khali, or "Empty Quarter." Swarms of locusts periodically emerged from those shifting dunes "long-legged in wavering flight," wrote Thesiger, "as thick in the air as snowflakes in a storm." He was hired to search for their breeding grounds, for the swarms regularly threatened the Middle East with famine, but he didn't take the job for entirely altruistic reasons. "I was not really interested in locusts," he admitted, "but they provided me with the golden key to Arabia." In the company of camels and Bedu guides, he spent months crossing the Empty Quarter, battling thirst and sandstorms. "To others my journey would have little importance," he acknowledged. "It would produce nothing except a rather inaccurate map which no one was ever likely to use. It was a personal experience, and the reward had been a drink of clean, nearly tasteless water. I was content with that."

After a week of biking through Oz, I would've been content with marginally potable water, but even that was scarce. We pedalled slowly in the gangrenous light of late afternoon, which gave the landscape the look of sunburnt, peeling skin. The profound silence of the desert didn't invite equally profound thoughts, but only drew my attention to the lack of sloshing in our water bottles and drom bags, now as empty as the Aral Sea. Mel became convinced she saw a building ahead, and I was equally convinced she was delusional until we biked right up to it.

Outside was a cot with no mattress, all wire and coiled springs, not unlike the gaunt man sprawled upon it. He blinked when we asked for *su*, or water, perhaps unconvinced we weren't mirages ourselves, then led us to some blue barrels that were bone dry. He seemed mildly surprised by this and beckoned for us to follow him to a rusty pipe that was dribbling water—and with it the smell of rotten eggs. We filled our bottles and drom bags as he watched, though we had no intention of drinking the gassy liquid unless truly desperate.

As Mel and I were preparing to bike away, a transport truck pulled into the driveway and an obese man slid out of the driver's seat, quivering wetly with perspiration. The thin man invited him in for a meal and extended the invitation to us as well. We followed him into a dark, airless room where more heavy-set, slack-eyed men, presumably truck drivers, sat at a table and tore gristly meat from bones plucked from a communal bowl. Their fingers left greasy smudges on the shot glasses from which they tossed back what I presume was vodka. Mel and I sat down in the chairs they pulled out for us, glanced uneasily at each other, excused ourselves, then fled.

Nothing like a room reeking of mutton fat and vodka fumes to make an oasis of the burning desert. We counted off the miles to Nukus, the westernmost city of any size in Uzbekistan, by coining neologisms for thirst, among them *aghh, wagh, grak,* and *mrwak.* "All our words sound like croaks," I pointed out to Mel. "It's a tonal language," she said.

Clearly they should've sent poets down the Silk Road, though I doubt even Don Domanski could have articulated the sweetness of the tea we drank at a roadside restaurant on the outskirts of the city. The tea's volcanic temperature had a paradoxical cooling effect, and the breeze peeled the sweat off our skin as it sent the poplars swaying. The trees lined the road in orderly ranks, their lower trunks painted white with some kind of insecticide. Next to them were irrigation ditches gagged with algae. The viscous, emerald sheen of the water was dimpled in places by the noses of frogs, looking like a slightly darker shade of algae, only the kind that croaks, so that even as I sipped tea all I heard was *thirst, thirst, thirst.*

The mother of four who ran the restaurant invited Mel and me to spend the night and even offered to wash our hair. By hand. With a bucket. Hair that hadn't seen anything but dust and sweat and the inside of a helmet for more than a week. Of all the hospitable gestures we were met with along the Silk Road, and there were multitudes, this woman whistling softly to herself as mud streamed off our scalps ranks among the most generous and loving. It felt so good to be taken care of, to let our guard down completely and be mothered like that. Mel and I were shocked to learn she was younger than us by a few years.

Before sunrise the next morning the broad avenues into Nukus bustled with people. Women strolled into the city wearing bright pyjama-patterned robes that clashed charmingly with their headscarves, and children skipped at their sides. Teenage boys in crisp blue jeans rode creaking bicycles past ours. Leather-faced men led donkeys whose soft noses almost scraped the ground as they hauled carts loaded with hay, tree branches, and in one case, three cows. In the dusty, slanted half-light of dawn the busy scene looked surreal and ghostly, a world of afterimages, as though I'd been staring into the sun too long.

Then again, I always felt this way re-entering the so-called civilized world. Many praise the courage and endurance of explorers, but what they don't realize is that some oddballs find routine far more terrifying than risk. How else to explain why Ernest Shackleton planned a human traverse of Antarctica, lost his ship to pack ice, spent years eating seal blubber while waiting to be rescued, and finally was rescued only to return a few years later to Antarctica? Or why Meriwether Lewis suffered all kinds of hardship as he and William Clark followed Sacajawea across the American continent to the Pacific Ocean and back, but when he was appointed governor of Louisiana Territory as reward for his achievement, he found civilized life so insufferable that he reportedly committed suicide? Or why Thesiger was drawn back to the Arabian desert over and over again, "for this cruel land," he said, "casts a spell no temperate clime can match"?

Far more torturous than melting on a bike in the desert, in our case, was shopping for instant noodles in Nukus. It didn't help that our clothes were hung out to dry after being laundered, leaving us with no sartorial options but to baste in fleece pants and long-sleeved wool shirts. I felt nostalgic for the air-conditioned supermarkets of America, where Mel and I had sought refuge from the simpering heat of our cross-country bike ride by filling shopping carts with all the groceries we dreamed of buying—watermelons and frozen pizzas, six-packs of Coca-Cola, diapers to cushion our saddle sores. This was a ploy to look busy long enough to cool down, at which point we unloaded the cart and walked out empty-handed. Shops in Nukus lacked air conditioning, but at least the hotel had cold showers and a manager easily bribed into back-dating OVIR receipts for us. On paper, Mel and I had stayed in the city every night since crossing the border into Uzbekistan. In truth, twenty-four hours in Nukus was more than enough.

The next morning we hit the road again: such meaningless penance, such profound relief. "All explorers must die of heartbreak," claimed the poet Charles Wright, but this mostly seemed true of those who tried to resume a normal life. The key, apparently, is to *never stop exploring.* Just ask nineteenth-century British naturalist Alfred Russel Wallace, who independently stumbled on natural selection as the mechanism for evolution before Darwin published his findings. But where Darwin retreated to his countryside cottage and never travelled anywhere ever again, Wallace never really settled down, either geographically or intellectually, and never lost his sense of wonder.

In his youth Wallace read *The Voyage of the Beagle* and took to heart Darwin's concluding message that "nothing can be more improving to a young naturalist than a journey in distant countries." But unlike Darwin, Wallace was the eighth of nine children born to relatively impecunious British parents who couldn't afford to subsidize his expeditions. Undaunted, Wallace struck out on the more entrepreneurial path of a freelance explorer, funding trips abroad by collecting exotic specimens from the Amazon and selling them to museums. Although he was no merchant like Marco Polo, this business model similarly gave Wallace the means and excuse to see the world, a way of life that suited him perfectly. To a letter from friends imploring him to return to England, he cheerfully responded, "Your ingenious arguments to persuade me to come home are quite unconvincing. I have much to do yet before I can return with satisfaction of mind. Were I to leave now I should be ever regretful and unhappy." When Wallace finally set course for home after four years in the Amazon, the ship he sailed out of Brazil on sank, and he was rescued by another vessel that barely avoided shipwreck itself. "Fifty times since I left Pará have I vowed, if I once reached England, never to trust myself more on the oceans," he confessed to a friend. "But good resolutions soon fade." Soon after landing ashore, Wallace was already dreaming up his next expedition, this time to the constellation of islands in what is today Malaysia and Indonesia, where he would conceive of natural selection in the flush of a malarial fever.

Wallace was no landed gentleman, so he faced the mundane exigency of making financial ends meet. This no doubt played some part in luring him back to sea, for he had to recover the catastrophic losses of his specimen collections to shipwreck. But if money was Wallace's sole imperative, there are far simpler, safer ways to earn a living than as a freelance explorer. And if fame and glory were among his ambitions, he could've insisted on far more of the evolutionary spotlight for his co-discovery of natural selection. Yet it is Darwin who earned himself eponymy with evolution, who stars in high school biology textbooks in which Wallace is hardly a footnote if he's mentioned at all.

Some might pre-emptively extract the moral for this evolutionary parable here: Wallace fatally lacked focus, you could argue; he wanted for direction. He dissipated the heft and genius of his life with ten thousand dilettante tacks of his ship, and ultimately sailed nowhere in particular (though just about everywhere in general, covering six of seven continents before he died at ninety-one). But since when is the measure of a life immortal fame? The more I'd learned about Wallace at Oxford, the more he'd seemed the explorer worth admiring. Where Darwin's quietly despairing *Autobiography* describes his growing indifference to the landscapes and works of art that once brought him "exquisite delight," Wallace's own memoir, *My Life: A Record of Events and Opinions*, relates how his passion for wild species and places grew year after year "as ever new and beautiful, strange and even mysterious forms [were] continually met with." If the highest goal humans can achieve is amazement, as Goethe attested, then Wallace led the more enlightened life.

His generative sense of wonder seemed to come from a refusal to specialize, to cultivate singular expertise at the expense of soul. When Wallace looked at the world, I suspect he didn't see fences at all. Over the years his writings on social and economic matters outnumbered those on natural history, and he openly stated that he saw the fight for personal freedoms as more important than the study of science. In a lecture, Wallace argued that every member of society deserved "all the essentials of a healthy and happy life." And what were those essentials? Everything of which the depressed elder Darwin was deprived: "ample relaxation, adequate change of occupation, the means of enjoying the beauty and solace of nature on the one hand, and art and literature on the other."

I don't mean to blindly apotheosize Wallace, who also penned anti-vaccination polemics. Nobody is perfect, not even explorers, which doesn't mean they aren't worthy of a selective kind of worship. What impressed me most about Wallace was his conviction that science and technology aren't sufficient, that we need a system of ethics as strong as our curiosity, a sense of restraint equal to our restlessness. In 1909, not long after the Wright brothers' pioneering heavier-than-air flight, Wallace sensed the vulnerability of the world to weapons from above and advocated for an international treaty prohibiting flying machines from carrying instruments of destruction. "Surely, for this great and holy purpose," Wallace pleaded in a newspaper editorial, "the whole body of true womanhood and true manhood will unite." The sky is not the limit, as the saying goes, but Wallace implied that perhaps it should be. After all, the history of science and exploration is a stirring adventure narrative about mountains conquered, birds envied and mimicked, limits of all kinds left behind in the dust. The history of science and exploration is also an argument for holding back.

So is the dying, drying Aral Sea, but like the rivers that used to fill it—the Syr and Amu Darya, or the Jaxartes and Oxus to the ancient Greeks—Mel and I wouldn't quite get there. With less than a month on our Uzbek tourist visas, we barely had time to bike across the country at the gruelling rate of sixty miles a day, never mind detour hundreds more to ecological disasters. I'd regret this later, for when would I ever be closer to that sea, or what was left of it? But we were too short on time to bike there and too broke to go by other means. Instead we settled for a brief side trip to Khiva, an oasis city and Silk Road trading hub that formerly specialized in slaves.

To get there we crossed the Amu Darya, whose polluted shores were as foamy as the necks of donkeys straining to pull loads over the floating bridge that spanned it. What impressed me the most in Khiva weren't the baked-mud walls, thick as houses, or the exquisite shades of the turquoise-tiled madrassas (medieval colleges for Islamic instruction), or the sense of history rising in waves of heat from the rammed earth. I was captivated by the Nature Museum, whose sign featured a sliced-up watermelon, a dinosaur, and a tabby cat.

Inside a cramped room next to one of Khiva's more nondescript courtyards were neat arrays of pickled snakes and frogs in jars.

Spiders and beetles were impaled on corkboard displays, though several had slipped off or broken apart, reminding me of Wallace. Even as he collected individual specimens for museum display, he rebelled against a reductionist approach to understanding the world, one that severs mechanics from meaning, ideas from implications, our heads from our hearts. We're all encouraged to become, like Darwin, a kind of machine for grinding out one specific purpose—basically the antithesis of the sort of roving, generalist explorer I longed to be. But there's more at stake here than the elder naturalist's quiet despair or my loud unhappiness in the laboratory. This sort of blinkered specialization enables Soviet engineers to build canals to irrigate a desert and also lets them shrug off the consequences of their constructions. Ecological collapse in the Aral Sea? Social collapse in seaside communities? "Sorry," they can say, "that's not my jurisdiction."

What can we learn about anything in isolation? Only that labels leave out more than they let on. By calling something "marginal" we make it so, when in truth *weed* is just another word for wildflower, *wasteland* is another word for wilderness, and *slave*, in ancient Khiva, referred to a person like you or me. All through the city's mud-walled labyrinths were niches where slaves were showcased as recently as a century ago, and the displays of creatures in the museum didn't strike me as so different in terms of isolating and objectifying life, reducing a coherent living whole to components or exotic curiosities. Which meant I wasn't sure what to make of a large jar in the Nature Museum containing a human fetus in a bath of formaldehyde. A label identified the specimen as a "brianless [*sic*], bornless child."

All across Uzbekistan Mel and I christened the places we pitched our tent: Camp Sore Ass, Camp Sweat Stain, Camp Desert Rain,

Camp Can't Eat Oatmeal Anymore. Mel managed to choke down the nutritious gruel, but one day I woke up unable to stomach it. After five months of mornings that began with the slimy, lumpish paste, no amount of sugar, milk powder, or hunger could salvage oatmeal for me. Instead I drank instant coffee and set off hungry, which made me particularly prone to crankiness when passing cars and trucks blared their horns at us, and every road worker or person we biked by whistled or cheered. They were just being friendly, so I gamely smiled and waved until the requests for acknowledgement grew so frequent, and the road so rough due to construction work, that I didn't dare risk releasing my death grip on the handlebars.

"We can't wave in response to every honk, whistle or shout," I grumbled to Mel on a rest break, expecting her righteous solidarity.

"Yes, we can," she said sullenly.

I stared at my friend. Her face was a froth of dust and sunscreen. Sweat beaded her skin as though she were being boiled alive. We sat in truculent silence as cars and trucks drove by honking excitedly. Neither of us waved. I tried to prop my arms on my bare knees but they slid off, frictionless with sunscreen and sweat.

"Well, why don't you then? Wave at everyone?" I finally asked, at a loss for what else to say.

A long, defiant pause. "Because I get tired."

Central Asia, where everything was fried: the land, the two of us, and also, thank god, the food. What saved our sanity, appetites, and possibly friendship were the roadside truck stops with deliciously greasy menus, which appeared more frequently the farther east we travelled in Uzbekistan. Instead of melting in a tent every afternoon, we napped on raised eating platforms that Mel fondly dubbed "playpens." I tried not to think about how many people had sweated into the faux velvet pillows and blankets covering them, even as I sweated into them myself. We assuaged our guilt over squatting there all day by ordering plate after plate of fried eggs or tomato-and-cucumber salads, each of which came with a cyclonic, buzzing side of flies. I wished I could twitch isolated precincts of my skin to flick them off, as horses do. Mel dodged them by doing post-meal yoga in the playpen. "Gymnastics!" one older man declared approvingly as he passed.

Despite the flies, the breezy shade sure beat the Glow-worm. I lay on my back and admired the upside-down world. Clouds traced ideograms of loneliness in the sky. Light was rerouted through the trees, their branches sharp detours for sunshine. Some birds tried to build a nest but the wind stole blades of grass from their beaks before they could place them. To pass the time I listened to "The Swimmer" by John Cheever as read on *The New Yorker* fiction podcast, not realizing its descriptions of a man on a swimming pool pilgrimage would be as torturous as biking into headwinds across Uzbekistan. "The day was lovely, and that he lived in a world so generously supplied with water seemed like a clemency, a beneficence." I almost wept.

Only in Bukhara, a few days later, did I encounter anything like the pale green pool fed by an artesian well in which Cheever's protagonist begins his quest. Pools throughout that museum of a Silk Road city looked similar at night, though not from algae, or a high iron content, but green spotlights inexplicably set up around the Old Town, giving its ancient mud walls a brash new vitality. At dusk the swifts reeled through those beams and turned briefly jade, and small boys on enviably unburdened bicycles cast little green shadow puppets on walls that mirrored and distorted their wheelies and skid-stop turns. One kid narrowly avoided riding into a canal that glowed as though radioactive, reminding me of the "arteries of sustenance" that Percival Lowell thought he saw on Mars early in the twentieth century. The American astronomer confused a skein of long, straight lines on the red planet (which don't actually exist) for an elaborate system of canals built by Martians to shunt water from the frozen poles to equatorial cities. Lowell even summoned Darwinian theory to support the existence of aliens on the red planet, arguing that intelligent beings had evolved once on Earth, so why not on neighbouring Mars? Wallace disagreed, so much so that he wrote a book refuting Lowell's claims, titled *Is Mars Habitable?* He argued that Mars was too cold, dry, and inhospitable to host an advanced civilization, and even if little green men (and women) somehow did exist there, why would they build such an impractical canal system in which vast volumes of water are lost to evaporation? The same could be asked of irrigation in Uzbekistan. Why, indeed, would sentient beings build such an impractical canal system to drain a living sea and grow one of the world's thirstiest crops in a desert?

I agreed with Sagan's conviction that "if it is just us, it seems like an awful waste of space." I also suspected extraterrestrial civilizations hadn't made contact because they'd decided, after eons of careful observation, that our planet is utterly lacking in intelligent life. Mel and I personally validated that conclusion when we set off biking again in the blazing inferno that was Uzbekistan. When we came across a concrete-lined canal we lay down in it fully dressed, still wearing our helmets, not caring that the water was a chemical slurry of fertilizers and pesticides, and ignoring the fact that on its paved shore a mangy, desiccated cowhide was still attached to its skull like a life-size poison label.

At least Samarkand's magnificent domes and turrets testified to human intelligence and ingenuity, if you could overlook the fact that they were built by the slaves of Timurlane, genocidal heir to Genghis Khan. A crippled shoulder and knee didn't stop Timur the Lame, as his name was derived, from building an empire that spanned Turkey to India in the fourteenth century—severing, in the process, former intercontinental trade routes between Europe and Asia, which put an end to the glory days of the Silk Road. Credited with murdering millions and constructing pyramids with their skulls, this warrior-nomad graciously spared artists, weavers, glass-blowers, writers, and other craftspeople in the cities he conquered so that they might glorify his capital in Samarkand, which, inarguably, they did. That Islam Karimov fancies himself a Little Timur says all we need to know about the president of Uzbekistan's leadership style. Timur himself, who wasn't Uzbek but Turco-Mongolian, has been co-opted as the mythical founder of the Uzbek nation. Never mind that modern Uzbekistan was created whole cloth by the Soviets in the 1920s, and that both Bukhara and Samarkand are traditionally Tajik cities stranded there because of the arbitrariness of Soviet frontiers.

The question of borders is a source of ongoing grief for tourism-starved Tajikistan, because of all the historic Silk Road sites in Central Asia, Bukhara and Samarkand attract the most visitors. Mel and I saw more foreigners in a few minutes in these two cities than during our previous six months on the Silk Road, though most pretended not to see us. The mostly older, mostly French visitors refused to meet our eyes as they looked around distractedly for their husbands or wives, who were forever ducking into carpet shops. Such tourists were usually identifiable by their scrupulously clean trekking pants, whereas bicycle travellers were instantly recognizable by their raccoon-eye sunglass tans, tattered clothes, and uncertain, lurching strides, like astronauts relearning how to walk. At least that's how I hobbled around Samarkand, marvelling at how its tiles and curves and tessellations suggested infinite visions, infinite viewpoints, even as the city was built (and rebuilt) to honour one kind of god, one empirical narrative. It seemed to me that maps should imply that same sort of limitlessness, for where you see one road and one country there are actually many, revealing themselves a little differently to everyone who travels them. Nations borrowed from the truth and returned by morning.

Doves cooed in the courtyard of Samarkand's Registan, where three peerless madrassas stare each other down across the centuries, and grass sprouts from smooth, bald domes of turquoise tiles. On a street nearby a group of skull-capped men held the corners of a sheet taut beneath a mulberry tree. High in it a young boy shook the branches to send the berries raining down, and much pointing and debate ensued over the specific limbs the boy should shake to yield the sweetest storm. The men shuffled back and forth with the sheet to catch the white berries as people have probably done for thousands of years in Samarkand, for mulberry trees have been cultivated here since the fourth century BCE, their leaves being the preferred fodder of silkworms. The men gently shook the harvest off the sheet into bowls, where the mulberries looked disgusting, like mashed grubs or silkworm pupae. The latter have been proposed as a compact, fast-growing source of protein for long-term space missions—which seemed reason enough to stay on Earth. I hesitated when the young boy offered me a white berry to sample, but fortunately it tasted less larval than it looked.

We saw more and more mulberry trees on the way to Tashkent, the Uzbek capital, for the country grew lusher by the mile. Crops, animals, or human communities sprouted from every speck of arable land, making it harder to find places to camp. One night we stopped in a field busy with bent-over workers, who shrugged when we asked if we could pitch our tent. Taking this as permission granted, we searched among rows of crops until we found a clearing, and then set up camp under the supervision of a dozen giggling boys and girls who'd followed us there in a lengthening line. Pitching the tent briefly offered them some entertainment, but when we started boiling water for instant noodles the kids grew bored and wandered off.

The fields were honeyed with light as the sun set, and they gave off the warm smell of hay. When it grew dark after dinner Mel and I tucked into the tent, though it was still too hot for sleeping bags. I lay on top of mine and was just about to drift off when I thought I heard footsteps. Instantly wide awake, I listened but could only hear the rasp of crickets, cars on the road, dogs barking in the distance, the low bluesy moans of cows—all the usual sounds we collectively call silence. Then I heard more footsteps, followed by hushed whispers. I nudged Mel and sensed her sudden alertness. We waited, the darkness as loaded with possibility as water gathering to a drop. Nothing . . . still nothing . . . then a riot of monkey whoops, cow moos, dog barks, and wolf howls punctuated by giggles.

We stayed quiet until Mel, as usual, couldn't contain herself. "BEEF AND NOODLES!" she hollered randomly, prompting the kids to scream in delight. They ran off laughing and shouting "beefandnoodles, beefandnoodles," repeating the foreign syllables like a mantra or spell all the way home, or at least until we couldn't hear them anymore, in part because we were laughing so much ourselves. Just another night on the Silk Road, with silence settling over the fields and the crickets resuming their own strange incantations, spells that conjured beads of dew from blades of grass and lulled us to sleep under a smoke of stars.

9.

THE SOURCE OF A RIVER

Pamir Knot

*F*or all Marco Polo's exaggerations about the Silk Road, he deserves credit for downplaying the hardships of his journey. When the Venetian merchant's caravan was raided by thieves in a Persian desert, he reported without hyperbole, in the third person, that "Messer Marco himself was almost caught by these people in that darkness." He managed to escape, but his companions were killed or sold into slavery. He was similarly tight-lipped about suffering illness on the Silk Road. In Badakhshan, where modern Tajikistan and Afghanistan meet, Polo noted—again in the third person—that "he remained sick for about a year."

I was far less stoic when I woke up feverish in Tashkent. "I think I'm dying," I rasped to Mel in the Uzbek capital, sounding as though I'd gargled with barbed wire. I suspected this was the outcome of waiting in line at the Chinese embassy a few days earlier behind a man who periodically let loose violent explosions of phlegm. At the time, it seemed a small price to pay for Chinese tourist visas, for once obtained the only remaining bureaucratic hurdle we'd face on the Silk Road would be sneaking illegally into Tibet again. That is, if we made it out of Uzbekistan, which hinged on the authorities accepting our fake OVIR receipts. I couldn't tell where my fever ended and the world began as we biked to the border. When the Uzbek officials proved more interested in our marital status than our OVIR receipts, I was too wiped to feel relief.

In Tajikistan we stopped at the first house we saw, hoping to camp in the yard. A clean-shaven man with nut-brown eyes came to the door of the whitewashed plaster building and introduced himself as Bobo. Despite the heat he looked crisp and immaculate in a white skull cap, a white T-shirt with the sleeves cut off, and white capri-style pants. His sandals slapped loudly as he led us to a grassy plot of land where we could set up the tent. I was desperate to crawl inside, but Bobo invited us back to his house for a meal, and I couldn't refuse because I could barely talk.

Bobo quietly repeated our names as we walked: "Katerina, Melissa, Katerina, Melissa." Inside his refreshingly cool, dark home, his wife laid out two cushions next to bread stacked like plates. Then she brought out bowls I was sure contained some kind of gristly meat, judging by the amorphous shapes inside them. When I raised a fork to my mouth, I tasted the cool freshness of tomatoes and cucumber instead. Just as I thought the meal was over, his wife entered the room with the main course: a pyramid of *plov*, an oily mixture of rice, meat, and carrots that both Tajikistan and Uzbekistan claim as their national dish. After a few token bites I retreated to my sleeping bag, where I should've been all day.

The next morning I tried to convince myself that my aching muscles, raw throat, and pounding head were a dramatic improvement from the day before. After an hour on the bike I gave up. I lay under a tree on my Therm-a-Rest mattress, choking back tears as sweat dripped across my skin, sometimes feeling like ants were crawling over me, and sometimes, to my horror, there actually were ants. An hour or two later Mel urged me back into the saddle long enough to reach the nearest Tajik homestead. Another family graciously took us in, but their home was so loud and hot and swarming with flies I almost would've preferred the ants. A fan stirred the heat around the room, in which an older woman, possibly blind, idly waved a tennis-racket fly zapper around. It threw off sparks as bugs soldered their bodies to the air. This stochastic noise kept me awake despite my exhaustion, as did the TV, and when that was turned off the dogs barking outside were just as bad. All night I took turns dousing myself fully-clothed with a hose in the yard and then dozing in the brief coolness of the water evaporating, until the heat woke me up again, so I'd shuffle outside to repeat the ritual.

When I was still sick the next morning, Mel arranged for us to share a jeep ride to Dushanbe, the capital city of Tajikistan, where I could see a doctor who spoke English. I don't remember much about that drive except for techno music so loud that each throb of the bass was a hammer to my skull. Somehow the old man next to me slept through it, his head swinging at me like a wrecking ball on the rough road, whose switchbacks were decorated with smashed cars and trucks.

The doctor's English was less fluent than I'd hoped. He peered down my throat, a puzzled look on his face, then scribbled something on a piece of paper. "I writing good recipe for you!" he said, not sounding very convinced. The pharmacist translated the prescription into four containers of liquids and pills that I was instructed to take at various times a day for the next week. At the hostel in Dushanbe, where we pitched the tent in the yard because it was cheaper than a room, Mel Googled the labels on the medicine to make sure they wouldn't hurt more than help. Her searches revealed that silver nitrate is seldom used in the Western world, for it can cause severe gastroenteritis that may end fatally, and that Dexoral is only approved for veterinary use in most countries. Cipfast was ciprofloxacin, more commonly known as cipro, a general antibiotic used to treat a wide range of bacterial infections. She couldn't find any information at all on Traclysan. I took cipro, put the rest of the good recipe aside, and went to sleep for what felt like a week.

When I finally crawled out of the tent I felt well enough to interview government ministers about wilderness conservation in Tajikistan. This was the main reason we'd planned to come to Dushanbe, and if anything, these meetings further aided my recovery, for they involved a quantity of waiting as recuperative as bed rest. Mel and I would schedule a meeting for, say, two p.m., and the government minister would graciously promise to send a car to pick us up. When no car appeared by three p.m., we'd call to inquire whether plans had changed, only to be told to expect a ride "soon." We'd wait another hour, call again, and be told, "Soon, soon." An hour later, the car would arrive and the driver would reassure us "in two minutes office." Twenty minutes later, after numerous inscrutable stops and starts along the tree-lined streets of Dushanbe, we would finally arrive at the minister's office only to be served tea and told to wait some more.

We quickly learned to bring our e-readers on these outings and I passed the time revisiting Rumi, who was born on the edge of Tajikistan and Afghanistan, though he also spent time in Iran and of course settled in Turkey. All these countries claim the poet as their own, which is ironic given that Rumi's whole oeuvre is one long argument against borders, as well as a radical call for the renouncement of power, wealth, and other things nation states generally hold dear. "Which is worth more, a crowd of thousands, or your own genuine solitude?" Rumi asks. "Freedom, or power over an entire nation?" I doubted the dictatorial president of Tajikistan had actually read these or any of Rumi's lines, for in perverse commemoration of the poet's eight hundredth birthday the Tajik government issued a coin with Rumi's face on it.

Perverse or perhaps appropriate, as Tajikistan is the poorest of all the former states of the Soviet Union. This country seemingly boasts more goats than people and more vertical than horizontal land, and more than half its gross domestic product derives from citizens working abroad and sending money home. Given the lack of funding for basic services in Tajikistan, I'd been pleasantly surprised that a colossal 17 per cent of the country is protected within Tajik National Park. Sadly, this statistic is less impressive than it sounds. Without money to hire rangers to efficiently patrol and protect the reserve, its mostly unmonitored terrain makes it an ideal conduit for drug trafficking and the illegal hunting of wildlife—including the curlicue-horned Marco Polo sheep and the elusive snow leopard. Mel asked one government minister roughly how much wildlife was poached from the park each year, and he said something tight-lipped to the translator.

"No poaching, or no statistic—it's hard to say," she offered grimly.

Though Tajikistan apparently lacks the funding to build robust fences or enforce fines in national parks, the government mysteriously conjured millions of dollars to build such critical infrastructure as the world's tallest flagpole (demoted a few years later to the world's *second* tallest flagpole, after Saudi Arabia built one twenty feet higher).

"Our tradition is optimism. It is Tajik way," explained the translator when I brought up such excesses, prudently refraining from translating my questions.

"Is he an optimist?" I asked, nodding toward the government minister. He was frowning into his tea and tapping his foot impatiently.

"No, he is more like Kazakh," the translator admitted. "Pessimist."

But pessimism seemed a decidedly Tajik trait whenever we mentioned our plans to bike across the Pamir Plateau. "It is very extremely there," a different government minister warned us. "*Very* extremely. Those mountains, you cannot breathe." Three of the greatest mountain ranges on Earth—the Hindu Kush, the Karakoram, and the Pamir—meet in what geographers call the Pamir Knot and locals call the Bam-i-Dunya, Persian for "roof of the world." Similarities with the Tibetan Plateau are more than nomenclatural, given the lowest valleys loom higher than most peaks in North America, and streams film over with ice even in the summer. But one person's extreme is another's comfortable, just as wildness is relative, and perhaps sanity too. Pedalling across deserts and mountains for months on end through varying intensities of rain, shine, hail, and snow? Reasonable. Chemically analyzing the microbial equivalent of cholesterol in a sterile laboratory six storeys high? Very extremely.

Clearly I was feeling better. When we biked out of Dushanbe a few days later I was charmed by everything: the crazy traffic, the potholes, the way car horns in Tajikistan could sound like police sirens, donkey brays, bicycle bells, trumpets, the simple *bleep* of a curse word redacted on TV. The road out of the city was half-melted, sticky as tar, but I felt so regenerated after the long break I was sure I could power through superglue. That lasted about an hour. As the sound of car horns faded the farther we rode from the city, so too did my energy. We took a break in a village and quickly drew a crowd of little boys.

They were more interested in our bikes than us. After talking expertly among themselves about the size and number of the gears and the quantity of panniers (or so I guessed from what the boys were pointing at), they grew concerned that passing cars would knock the bikes over, parked as they were on the road shoulder. With our permission they wheeled the bikes to a more protected spot, taking enormous care to leave them exactly as they'd found them, right down to the flourish of Mel's backpack propped against the rear wheel. A few hours later, during another break, a different group of small boys threw stones into the branches of a nearby tree to knock down apricots, which they gallantly offered to us. The fruit was delicious, possibly because it was forbidden: while some boys raided the tree a few others kept watch for the orchard's owner. Eventually the boys grew tired of throwing stones at apricots and started aiming them at each other. We distracted them by letting them try on our helmets, which they loved because they could whack each other on the head and not feel a thing.

The road went from paved to gravel to paved again. The heat went from mean to murderous. At one point we crossed a bridge on which a pipe had burst, so that cold, clear water arced everywhere, spraying the road and the two of us as we biked through it. This impromptu shower kept us going to the top of a pass, where we could see the full scope of the Nurek Reservoir. It looked like a bathtub of turquoise water drained lower than usual, exposing a rim of bright red soil, like a rust stain. Created by a Soviet-era dam on the Vakhsh River, this reservoir generates hydroelectricity and also irrigates local farmland, and another, even bigger, power plant is under construction on the same river—to the chagrin of Uzbekistan, which depends on that water downstream. Tajikistan's true wealth is gravity and water: the Pamir Mountains harvest rain and snow and store it in glaciers, which melt into rivers, which serve as the arteries of life in parched Central Asia. Among these rivers is the Pyanj, known historically as the Oxus and farther downstream as the Amu Darya, which we'd crossed on a bridge in Uzbekistan more than a month ago.

The only place that river still gushes into the Aral Sea is on old maps, or at least the Silk Road map we carried, which hadn't been updated to reflect the modern, less liquid, reality. In the tent that night I pored over those contours, following the river's blue swerve with my finger from the Uzbek desert into Tajikistan, where a road ran parallel to it, beginning in the eastern foothills and wending up and across the Pamir Plateau. Along its course the river marks the border between Tajikistan and Afghanistan, and from what I could tell on the map, a high-altitude lake called Zorkul eventually swallowed it whole—and with it, the only visual evidence of the Tajik-Afghan divide. This river, in other words, was a chance to trace a border to its source.

It took us another week to reach it. We biked past rolling green hills like scoops of mint ice cream, past a donkey waiting in the shade of an old Soviet bus shelter. For once in Central Asia, we weren't in a rush: Tajikistan had generously granted us sixty-day tourist visas, enough time to bike across the country twice, which meant we could take the Silk Road as it came. And in the western half of Tajikistan, it came slow, hot, and rarely horizontal. After a grinding climb into cooler layers of sky, we biked down past cliffs so red it seemed the rock was bleeding, and only then did we reunite with the fabled Oxus.

Beneath our wheels was Tajikistan, on the far bank was Afghanistan, and all around us mountains rose like cupped hands, the river running from them like an oblation. Ragged peaks spliced the sun's general shine into neat rays, precise beams, illuminating one swath of the world and then another. Biking along the Pyanj was like going from black and white into colour. For hours and hours we'd travel a stark and seemingly lifeless land, a stubble of rock, and then see it suddenly bloom green where a creek silvered down the slopes. Villages took root alongside this vegetation. Stone buildings and stone walls surrounded neat mosaics of wheat and barley fields, cherry and apricot orchards, and groves of mulberry trees, the latter prized here less as fodder for silkworms than for the sustenance of the berries themselves.

The mountains above these villages were generally taller in Afghanistan, also sharper, like the nicked edge of a sword—the kind of landscape you could run a finger along and draw blood. Yet we saw more evidence of strife on the Tajik side. Military watch-towers tilted on struts above the Pyanj, and rusty green tanks littered its bank. Signs along the road showed cartoon human legs being blown off, warning of the minefields left over from Tajikistan's civil war in the wake of Soviet independence. We stayed on the road to avoid land mines, hesitant to even pee off its shoulder, and camped in the yards of families for the same reason, but the most hazardous aspect of Tajikistan proved to be the weather. One afternoon a storm blew in while Mel and I were on an exposed stretch of road, so we ducked under an overhanging cliff to wait it out. Thunder and lightning lashed the mountains, and hail the size of molars hit the metal guardrail, with a sound like teeth chattering. The hailstorm thawed into heavy rain, which loosened the dirt stabilizing the slopes above us and sent rocks bombing onto the road. Once the storm had passed and the world seemed calm and stable again, we got back on the road and started pedalling. Seconds later, a rock the size of a human head fell next to Mel with a sickening thud.

One afternoon we saw an idyllic campsite tucked on a hillside behind a low stone wall. It was shaded by trees, with a well-worn trail leading to it. Assuming this obvious path meant it was safe from land mines, we hiked up the hill and stretched out behind the wall for a nap, figuring we'd wait until dark so as to pitch our camp unseen. An hour or so later we heard voices. I peered cautiously over the wall: on the road were six or seven men wearing camouflage and carrying guns.

"Should we let them know we're here?" I whispered to Mel, not wanting to surprise a Tajik military patrol. But it was too late. A soldier spotted our bike tracks in the dirt of the path and followed it with his eyes up to the wall where we were spying on him. Mel and I quickly stood up and shouted hello, waving our hands in what we hoped was a friendly fashion. He motioned for us to come down.

The soldiers were alarmingly young, tall, and gawky. Using hand gestures we explained we hoped to camp here. At first they nodded in agreement, but after talking among themselves, they changed their minds. One of them pointed at Afghanistan and mimicked someone swimming across the river and then aiming a gun. The bank was barely a stone's throw away, but the river frothed and churned with such force it was hard to imagine anyone making it across. The soldiers insisted we return to a guest house in the town we'd biked through earlier, now two miles behind us and up a horrible, hilly stretch of gravel road.

I'd rather bike ten miles forward than two in reverse, but the sun was setting and we had no choice. The soldiers watched us as we packed up our bikes, wheeled them down the path, and set off grinding up the first steep hill. When we reached the village, we were told there was no guest house, but a family of sweet if slightly off-kilter women took us in. One daughter, who was roughly our age, giggled constantly and was otherwise mute. Another, slightly older daughter had one blind, opaque eye and never smiled. We helped them milk some yaks, meaning Mel and I tugged uselessly at the teats to the entertainment of the gathered bystanders until the daughters took over, expertly sending white spurts into a pail.

Inside their home, crimson carpets with floral patterns covered not just the floor but also the walls, as if you could walk right up them. Round tree-trunk beams in the ceiling were painted turquoise, with rough-hewn boards in between. A flickering bare light bulb dangled from a beam, and plants that looked tenderly cared for flourished on the windowsill. A boxy metal stove was balanced rather precariously on rocks and wood, an occasional flame escaping from its hinges. When the mother—an ample, exuberant woman whose breasts practically sagged into her pockets—put a kettle on the stove, I snuck a peek under a floor carpet: linoleum the colour of wheat.

As we ate homemade bread with butter and fresh cucumbers for dinner, the mother carefully wrote down her cellphone number on a scrap of paper and gave it to us. I tore a page from my journal and gave ours to her in turn, though there was no reception in that part of Tajikistan, and even if there had been, neither of us could understand a word the other said.

Centuries ago, both banks of the Pyanj River were part of the same political territory known as Badakhshan, which was populated by Ismaili Muslims and ruled by various emirates. The boundary now severing the region was drawn in the Great Game, a British-Russian standoff over Central and Southern Asian territories during the late nineteenth century. The Emirate of Kabul, supervised by British India, ceded the east bank of the river to the Emirate of Bukhara, a Russian protectorate. Although local trade and travel took place across the Pyanj for nearly a century after the border was established, the situation changed following the Soviet invasion of Afghanistan in 1979, the rise of the Taliban in the 1990s, and the civil war beginning in 1992 in Tajikistan, with the boundary growing more militarized and less fluid. Families found themselves stranded on opposite banks, able to wash clothes in the same water but forbidden from boating across.

I felt the vertigo of that divide as Mel and I biked farther along it. The road in Tajikistan, though rough, was paved in places, and some of the families we stayed with watched television over dinner. Across the river in Afghanistan—seemingly across the centuries—the stone-hut villages went dark at night and there was no road, not even a euphemism for a road, just donkey tracks scuffed into the riverbank. Some mornings we saw Afghan girls in elegant indigo robes walking those paths, presumably to attend school in a nearby village. They glanced over at us with—curiosity? Alarm? Longing? Pity? Their faces were covered, so we couldn't tell.

Even for Tajiks the sight of foreigners on bicycles was unusual. In one town a young mother was bathing her child in a stream when he spotted us riding by. The boy was so excited he ran after us wearing nothing but sandals. I thought back to university in North Carolina, where a home team basketball victory would send students running half-naked into the streets to flip cars and leap over bonfires. I could never muster that kind of ecstasy over a sports win. For me to run nude and rapturous down the road, it would take something like NASA announcing the discovery of life on other worlds—which is basically what sent the little boy running after us. He stared at Mel and me, giggling and dumbstruck, until his mother came to retrieve him.

A few days later we were met with similar awe by two sisters, though they were fully clothed, in colourful patterned dresses that looked like extra-long shirts. These girls seemed to be between ten and twelve years old, one with brown curly hair, the other with straight black hair, and they were playing by the side of the road when we stopped briefly to say hello. They proceeded to pull out all the stops so that Mel and I would stay. First was a song and dance routine, in which the sisters warbled and twirled around until they were dizzy, which was clearly part of the fun. Next, they gave us a tour of the dollhouse they'd ingeniously built with bottles and bits of trash, a make-believe mansion on the edge of their yard. When we turned our backs on this work of art, two boys snuck over and stole pieces of it to throw meanly at the girls. Mel chased after them and squirted them with her water bottle, prompting the boys to run and howl in mock fear and the sisters to cheer. Then the girls and the two of us had a water fight ourselves, because it offered relief from the heat.

It was an afternoon of pure play, aimless and timeless and requiring no translation. The girls' mother offered for us to pitch our tent in their yard, which we did, much to the fascination of the girls. Once it was up they scurried inside and swept it out with a broom, then inflated the Therm-a-Rest mattresses and unfurled the sleeping bags for us. That evening the sisters tried on our helmets and sunglasses, looking more bad-ass in our cycling gear than we ever did. We propped the bikes on kickstands and helped the girls into the saddles, where they pretended to ride even as their feet dangled high above the pedals. They leaned from side to side through whiplash curves, squinting through dust kicked up by the wheels, and crouched low over the handlebars as though travelling at tremendous speed.

At lower altitudes, the Pyanj ambled along at the pace of someone with no particular place to be. Higher up, the river shed its load of silt and turned indigo as it steepened and narrowed, with Afghanistan inching closer and closer. Rapids undercut the road in places, dissolving slabs of gravel or concrete like salt, forcing us to swerve wide or be swept away. Ragged mountains rose steeply on all sides, so that the town of Khorog looked pinched in the jaws of a piranha. When we stopped there for dinner at a local Indian restaurant, a giant poster with our blown-up, pixelated pictures on it personally welcomed us to the Pamir—or rather, welcomed me and "Mellisa Yue."

A friend of a friend, Aziz Ali, had learned we were passing through Khorog and organized a party in our honour. Originally from Pakistan but now living in Afghanistan, where he works on community development in the Pamir borderlands with Tajikistan, Aziz Ali was a kind, swarthy man with round cheeks and a sing-song accent that made music of normally bland sentences. He gave a long welcome speech endearingly peppered with "wholly solely," as in, "We are gathered here tonight to celebrate these brave Canadian women who have biked here from Canada *wholly solely* by themselves!" If this wasn't enough to make us blush, we were gifted with bouquets of roses and gorgeous Pamiri necklaces, an intricate weave of red and white and black beads that made even our ratty T-shirts look stylish.

There were about nine or so people gathered at the dinner, and one of the men, sitting across from Mel, kept glancing back and forth between the welcome poster and my friend. "Yue is a very strange surname," he mused out loud. "It sounds Chinese, and yet you are not Chinese?"

Conversations around the table stopped. Everyone stared curiously at Mel, awaiting clarification.

"Well, haha . . . ," she hedged, struggling to find a delicate way to point out the spelling error. "You see, it's no big deal, but my last name is actually *Y-u-l-e* . . ."

Stricken looks all around. The truth has a time and a place, it turns out, and that's rarely on the Silk Road—a lesson further reinforced when a thirty-something ethnobotanist named Munira, also at the welcoming party, asked us how much our bikes had cost. We explained that our custom-built titanium touring bikes had actually been gifted to us by a generous bike company. So then she asked what they would have cost, if we'd had to buy them.

"Um, around fifty dollars?" I lied.

"Yeah, about that," Mel quickly agreed.

Munira's eyes widened. We later learned that even this vastly downplayed price tag was more than her monthly salary at the Pamir Biological Institute, which despite being one of Tajikistan's top scientific institutions has no Internet connection, not nearly enough lab space or offices to go around, no subscriptions to the latest scientific journals, no funding to send Munira and her colleagues to conferences to engage with the broader scientific community—none of the resources, in short, that I'd had in abundance at MIT.

I thought of our exchange a few days later, in the village of Darshai, where we stayed with a geography teacher whose lively, exuberant mind seemed too vast and searching for the narrow circumstances it found itself in. Though Mubarak Sho spoke some English, we couldn't understand what he was saying about dinner, so he drew a cartoon of a long-eared rabbit caught in a snare (Mel was by now a "Central Asian vegetarian," meaning she ate whatever she was offered). The meat—gamey, tough, but delicious—was served after an elaborate pre-feast of fresh bread, fried eggs, sweet milky chai, and a plate full of cookies and candies, for in Tajikistan dessert often comes before dinner. During the meal Mubarak revealed his intimate, far-ranging knowledge of the world by effectively giving himself a pop quiz and acing it. He listed Great Lakes in Canada like Superior and Erie, and rivers in the United States like the Mississippi and the Missouri. He described tribes in the Amazon jungle, mimicking blowpipe hunting to get his point across, and illustrated the Arctic tundra with cartoons of an igloo and a whale blubber lamp. Kangaroos, sharks, elephants, and cobras were discussed in turn, as were all the major mountaineering peaks in the Himalaya, from Everest to Lhotse to Annapurna.

How did he learn all of this, I asked him, living in a remote village in Tajikistan with, as far as I could tell, no library and no Internet? But Mubarak didn't understand my question, for while his English was impressive, it was apparently limited to the wonders of the world—which were everywhere, by his measure, except Tajikistan. "What is Zorkul Lake compared to Machu Picchu?" he exclaimed when he learned that's where we were headed. "Built by cosmos, Machu Picchu. By aliens."

Here Mubarak's knowledge fell short: when we arrived at Zorkul Lake a week later, I nearly fell off my bike at its beauty. The dark blue of the lake was liquid twilight, as though the sun were always setting on the Pamir Plateau and the stars were about to emerge from that water. The long, open valley in which the lake was set was surrounded by mountains: sharp, ragged summits on the Afghan side, and gravelly, rounded slopes in Tajikistan. Even the valley floor loomed fourteen thousand feet high, but I hardly noticed the lack of oxygen: the air was fresh and cool and full of unexpected fortunes, like hints of sage and glacial ice. The wind untied my shoelaces as I biked but at least it was coming from behind us, propelling us forward, though the boost was mostly wasted on the rough road. At times it disappeared completely. We'd find ourselves pedalling across lush carpets of grass scattered with yellow flowers, or splotches of purple lichen in perfect concentric circles. As long as we kept the lake to our right, we eventually relocated the road, or the faint trace parodying as one, which was my favourite kind anyway: barely distinguishable from a trail, or better yet barely there at all.

From Zorkul I could almost see the end of our Silk Road. Across the lake was the Wakhan Corridor, the narrow finger of Afghanistan that points to China between Tajikistan and Pakistan, where Siachen spills over into northern India. The glacier was just a few hundred miles away, or a few days of biking as the crow flies, but political borders meant it would take three more months to get there.

I thought about Fanny Bullock Workman, who didn't live to see her beloved "Rose" lose its status as the world's longest non-polar glacier (the Fedchenko in Tajikistan proved longer by less than a mile). Nor did she live to see Siachen turn into a war-torn garbage dump, which surely would've broken her heart. It broke mine, after all, and I'd never even laid eyes on its ice. What I had seen was the Juneau Icefield, which suddenly struck me as pivotal. Without that summer on ice, the saga on Siachen probably wouldn't have obsessed me so much. I wouldn't have been able to imagine a remote glacier in Kashmir and so wouldn't have lamented its desecration any way but abstractly. Instead, caring about one icy borderland had primed me to care for another that I'd never seen. Maybe exploration at its best is about building that kind of metaphorical muscle. After all, the term *metaphor* comes from the Greek *meta* (above) and *pherein* (to carry)—to be carried above, a flight into connection, so that after travelling long and far enough every mountain reminds you of another mountain, every river summons another river, and you learn enough landmarks by which to love the whole world.

So what happens when you can't travel, not in words, not in the world? Along the eastern shore of Zorkul we passed a rock hut chinked with dung outside of which yaks grazed. A little boy ran outside when he saw us, but it was getting too late in the day to stop, so Mel and I waved goodbye and continued biking. In response he threw a fistful of stones at us. The gesture seemed more full of mischief than malice, but we pedalled hard to get beyond arm's reach, just in case, and I often wondered whether that boy had darker, more ineffable targets in mind, such as the freedom certain humans, by total fluke, are born into, or the fact that the same road leads different people different places. Mel and I were just passing by, moving on, the wind erasing our tracks behind us.

Unlike political frontiers, so crisp and martial—precisely here is Tajikistan, exactly there Afghanistan—ecological borders are more often murky, a mosaic of give-and-take: the thinning of greenery above the treeline at Zorkul, say, or the interlude of dusk that drew marmots from their dens. The scientific term for such natural frontiers is *ecotone*, coined from the Greek *oikos* (home) and *tonos* (tension), suggesting that being truly rooted requires a certain restlessness, that home is a less static place than a state of potential energy. If it weren't for the rigid walls built by politics, the concept might apply to everyone living in the Pamir, from human communities to flocks of Marco Polo sheep. Not that I could see them: the herds blended in with the boulders so seamlessly that land and creature became one, sheep being the part of the mountains that moved.

"See? Many, many. See there," exclaimed Sergei, a weather-beaten guide we'd hired to help us spot the herds. But when I glanced through his spotting scope I only saw gravel, sky, clouds. Mel took a look and shook her head. The shy herd had disappeared over a ridge.

We got back into Sergei's 4WD jeep and he tried to drive closer, which meant gunning up a steep slope of loose dirt, gravel, and grass. As the wheels spun out uselessly Mel nodded in sympathy. "Story of my life," she commiserated. Sergei cheered in triumph when he finally managed to crest the hill, prompting Timurlane, his fourteen-year-old son, to roll his eyes as if his father were insufferably unhip. A city boy exiled to the Pamir for the summer, he wore a black faux leather jacket, black jeans, scuffed sneakers, metallic aviator sunglasses, and a glowering look that befitted someone named for Central Asia's deadliest ruler. This incarnation of Timurlane also had political ambitions, but fortunately they involved going to Oxford and becoming a diplomat. When I mentioned I'd gone to Oxford myself his face lit up until he remembered that unabashed enthusiasm wasn't cool, which was the story of *my* life.

Sergei stopped the car and waved toward a mountain slope that looked empty. I looked through the spotting scope in the direction he'd gestured, and what I'd taken for boulders resolved into hundreds upon hundreds of sheep. The flock poured over the land like light, at once particle and wave, moving up the mountain with a liquid grace that left me stunned.

"Can we go now?" said Timurlane, bored beyond belief. He kept playing ring tones on his fancy cellphone as if that might summon reception. His father could probably afford to send him to Oxford, not from guiding tourists like Mel and me, who wanted to shoot photos of wildlife, but from foreign trophy hunters who wanted to shoot guns. Although the Marco Polo sheep is a threatened species, foreigners can pay up to forty thousand dollars to hunt down a pair of enormous curlicue horns in Tajikistan. Ironically, those same horns were scattered across the plateau, the discards of wolf or snow leopard kills. "What the wolves leave behind," Sergei chuckled, kicking at an old, yellowing pair of horns, "foreigners hang on their walls."

And Tajik guides hang photographs of foreigners on *their* walls, at least judging by Jarty Gumbez, home to the private hunting conservancy Sergei worked for. Its wood-panelled library hosts a billiards table, a big screen TV with a satellite connection, and a shrine of images from successful trophy hunts: triumphant portraits of ruddy-faced men—I saw only one woman—in puffy white camouflage jackets posing next to their prizes. The wild sheep in the photos looked enormous until I realized that the hunters sat a short distance behind them, so that perspective tricked the trophy horns into titanic dimensions. In some images the ram's teeth were bared in an unsheeplike snarl; in others the ram looked restful, at peace, as though he'd just laid his magnificent head down for a nap.

A gunshot is a singular, stunning act of violence, which makes trophy hunting easy to condemn, especially when it isn't for the sake of sustenance, but an ego boost. What's harder to perceive, never mind prevent, is the less visible, more complicated violence of life in a country with limited options. Chronic poverty in Tajikistan means that wild sheep will be killed no matter what. It's just a question of who will pull the trigger: Soldiers in the country's corrupt military, who, Sergei told us, were the only people with guns in Tajikistan following the confiscation of weapons after the civil war? Locals who rent guns from soldiers because they are desperate for meat or the black-market money it brings? Or the occasional foreigner keen on snagging a fancy wall decoration and willing to pay a minor fortune for it?

Trophy hunting at least gives conservancies like Jarty Gumbez the incentive and the means to protect wild herds of Marco Polo sheep, if only to ensure the sustainability of their business model. A grim calculus, to be sure, but there are worse things than sacrificing an aging ram every few years to subsidize the health of the herd—and of local communities as well, for the conservancy generates a number of well-paid jobs. Of the estimated 23,000 Marco Polo sheep in Tajikistan, it's no fluke that nearly half live in the Jarty Gumbez conservancy, which does a better job at protecting wildlife than national parks. And when Marco Polo sheep thrive, so do the wolves and snow leopards that eat them, making foreign trophy hunters the unwitting heroes of ecological conservation in Tajikistan.

I peered at the wall of photographs. The hunters didn't look so heroic in their white camouflage suits, guns slung across their backs in dark slashes. They looked, it struck me suddenly, like soldiers on Siachen. And I wondered if the conflict in Kashmir was just a distraction, a flashy but misplaced target for my grief over the loss of wildness from the world, the way trophy hunting is easier to denounce than more subtle forms of violence in Tajikistan, such as the poverty that trophy hunting actually helps, in a roundabout way, to ease. After all, glaciers everywhere are vulnerable to the slow devastation of climate change, the war that extravagant lifestyles in North America and Europe are waging daily against ice. The world's highest and biggest garbage dump isn't a Himalayan glacier, but the atmosphere above it. In that sense we're all citizens of the same country, complicit and connected, and the more I stared at the photos, the less convinced I was about what I saw. The creaminess of the Marco Polo sheep wool was just slightly darker than the general whiteness into which they slumped, like shadows cast on snow, if only shadows could bleed. All around them were diffuse crimson smears the precise shade of the Juneau Icefield each summer, when snow algae blooms in pale red sheens across the light-soaked ice, making those glaciers briefly burn.

The hardest thresholds to cross are rarely as tangible and obvious as a fence. This seemed especially true in Tajikistan, where the only physical barrier Mel and I saw, as we pedalled north past the town of Murghab, was a dilapidated string of barbed wire along the Chinese border. It was missing so many wooden posts you could hop right over it, and judging by the tracks I saw, many humans and animals did. Originally built in the Soviet era to mark the edge of a broad buffer zone with China, this fence became the actual border between the two countries when Tajikistan ceded almost 1 per cent of its territory to China—akin to Canada giving away New Brunswick or the United States ceding Indiana—in order to settle a centuries-old territorial dispute. As far as fences went, rumour had it that other countries in the region planned to follow suit in the name of national security, which seemed absurd when the Pamir range, looming sharp and ragged as broken glass, prevented human trespass better than barbed wire ever could. Mountains and lakes and rivers are the oldest kinds of borders, and maybe the only sort I fully respect.

The frontier between Tajikistan and Kyrgyzstan was high on a mountain pass. As we approached it an English-language sign commanded STOP, and on it someone had scrawled, "It's hammer-time." The border compound looked as though someone had taken a hammer to the place. Dilapidated trailers sagged next to a building that was half-built or half-demolished—it was hard to tell—but either way had an apocalyptic air. A sleepy-looking Tajik soldier materialized from the gloom and stood beside us, inspecting our bikes. Mel's was propped upright but mine was on the ground, having jettisoned its kickstand a few days earlier. The soldier prodded my bike's back wheel over and over with his foot to keep it spinning in the air, and with each nudge the gun on his back wobbled slightly.

After a few minutes an official ushered us inside a dark trailer full of tables half-buried beneath yellowed forms. He gestured for us to sit down by patting the grubby mattress of a bunk bed. As he jotted down details from our passports, another soldier nearby was eating lunch. When he saw us watching he held out his bowl, generously offering to share. "What is it?" Mel inquired. The soldier grinned, then gripped his fingers into the shape of a gun, made a *bang*, and twirled his hands at his temples as if to convey craziness, someone off his rocker, and the Marco Polo sheep's trademark curlicue horns.

10.

A MOTE OF DUST SUSPENDED IN A SUNBEAM

Tarim Basin and Tibetan Plateau

As soon as Mel and I had a tentative grip on one country's language and customs, we left it behind, though in Kyrgyzstan's case not before a night in national limbo. Neither Tajikistan nor its neighbour apparently felt obliged to maintain the road between their border compounds, for it was a whiplashing plunge of ruts and loose rocks. Marmots sounded shrill alarms from the meadow we eventually camped in, and we expected similar alarms at the Kyrgyz border post the next morning, when the guards squinted suspiciously at our Kazakh visas amended with their country's name. To our relief they shrugged and stamped us in. If all went well, we'd be stamped out again within thirty-six hours. China was barely a day's ride away.

The now-paved road spat us into the Alai Valley, a lush sweep of grass dotted with yurts, horses, and herds of domestic sheep. The mountain range we'd emerged from that morning loomed higher the farther we biked away, an upheaval of white light. On the first and only night we camped in Kyrgyzstan, three teenagers galloped over on horseback, a dead sheep slung over one saddle. The sheep's throat was slit into a garish red smile and its legs flopped against the horse's ribs. The Kyrgyz boys—gangly, rail-thin, seared dark by the sun—were friendly and curious, lobbying us with questions and repeating them louder when we didn't understand, as though translation were a simple function of volume.

The next day Chinese border officials tried a similar strategy of amplifying decibels in their interrogations at customs, but Mel and I had forgotten most of the Mandarin we'd picked up five years earlier. As we waited for our bikes and panniers to scan through X-ray machines, I idly wondered whether Marco Polo had consulted a phrase book on China's Silk Road. They were certainly in circulation back in his day, and the similarities between twelfth-century editions and modern incarnations are striking. One Tibetan-Chinese edition I'd read about from Polo's era provided translations for food, clothing, and tools, as well as expressions for seeking a bed and meal in strange towns. It also included phrases helpful in dealing with common travel problems, such as illness, thievery, or being accused of a crime, including the timeless plea, "What have I done wrong?" In retrospect I wished I'd memorized the Mandarin for that one, for we needed it our first night in China.

Beyond the border was the trucker's pit stop of Simuhana, the westernmost town in the country and a place sometimes evoca-

tively referred to as "the last part of China the sun's rays touch."
When we arrived there, I suspected the sun had its reasons for delay.
The town boasted more of everything than similar-sized outposts
in Central Asia: people, noise, transport trucks, trash. Its streets
were an extravagant squalor. Plastic instant noodle wrappers fluttered in the breeze and crunched underfoot. The main drag was
paved in broken glass and the blood of a freshly butchered cow, its
carcass hissing with flies on the sidewalk.

Why we decided to celebrate our return to China by treating
ourselves to a guest house stay, I can't recall, given that whenever
we paid to sleep somewhere instead of pitching our tent for free,
we typically ended up with insomnia, whether from a hotel faucet
that wouldn't stop dripping (another form of water torture we'd
encountered in Turkey) or techno music throbbing through the
floorboards from the restaurant below that turned into a disco at
night (as had happened in China on our previous trip). But the
guest house we found in Simuhana seemed quiet enough. It was
run by a family of Uyghurs—pronounced wee-gerz—a Turkic
Muslim minority who suffer the same persecution in China as
Tibetans, only they lack the equivalent of a Dalai Lama rallying
sympathy for their plight abroad. We preferred supporting them
to stuffing more yuan in the pockets of Han Chinese, the country's
dominant ethnic group, who have been relocated en masse to the
Xinjiang Uyghur Autonomous Region in northwestern China
(just above Tibet) in a policy of calculated resettlement. We also
couldn't wait to eat laghman again, the signature dish of Uyghurs,
a spicy tangle of hand-pulled noodles and bell peppers, which we
were offered upon arrival at the guest house.

After dinner the teenage daughter of the guest house owners
and the two of us had just settled into watching Uyghur music
videos—kohl-eyed young men crooning on sand dunes—when

the room exploded into shouts and the stutter of camera flashes. Without a knock at the door, without a hint of courtesy or respect, certainly without a warrant, four uniformed Chinese police officers stormed in, hoping to catch us and our hosts in whatever criminal act we were suspected of: plotting Uyghur rebellion, say, or simply breathing the same air.

"PASSPORTS!" demanded someone I couldn't see because I was blinded by camera flashes.

"Janada, embassy, big problem!" Mel shouted in defiance. She then pretended to dial the Canadian embassy on our cellphone, though we hadn't bought Chinese SIM cards yet.

"PASSPORTS!" the police officer bellowed again, calling Mel's bluff. We never learned exactly what prompted this invasion, but from what we could gather, the guest house owners didn't have the correct permits for taking in foreigners as clients, which we didn't realize. Mel and I didn't want to surrender our passports, so we gave the police photocopies and agreed to visit the station in the morning, knowing that if we skipped town, our hosts, not us, would suffer the consequences. I was relieved when the officers left and the teen girl went back to the music videos, as though a police raid was just business as usual in Simuhana. Maybe in China, for Uyghurs, it was. Needless to say, I didn't sleep well that night.

At dawn we made our way to the police station. A sheep was getting butchered on the sidewalk where the cow had been yesterday. Uyghur women in long patterned dresses swept dust from the streets, a heroically pointless effort on streets made of compacted dust. Chinese men sent dark jets of phlegm flying from their doorsteps. They wore T-shirts hitched up over the hairless globes of their bellies, either a local style or a tactic to stay cool in the already fierce heat. I heard murmurs of *"polis,"* "passport," and *"Janada"* as we passed them. Mel and I signed our names to some kind of inscrutable statement, testifying to who-knows-what infringements, and skipped town.

As former American president John F. Kennedy liked to point out, the Chinese word for "crisis" is comprised of two characters, one representing "danger" and the other "opportunity." An astute and fascinating observation, except for the fact that it isn't true—which hasn't stopped management gurus, motivational speakers, and New Age pundits from touting it. The first character in the Chinese word for "crisis," or *wēijī*, does imply danger, but while the second appears in "opportunity" it doesn't signify it, just as the syllable *ex* doesn't automatically convey *explorer*. Instead, the *jī* character suggests "an incipient moment" or "a crucial turning point." "Thus, a *wēijī* is indeed a genuine crisis, a dangerous moment, a time when things start to go awry," explains Victor Mair, distinguished professor of Chinese literature. "It is not a juncture when one goes looking for advantages and benefits."

When Mel and I reached Kashgar, though, China itself seemed to have embraced the "crisis = danger + opportunity" mistranslation. This oasis city is where the northern and southern routes of the Silk Road meet after skirting the Taklamakan Desert, and it served as a crucial stopover for trade for thousands of years despite repeated sackings, first by Genghis Khan, then Timurlane, and most recently the Chinese government. When we first biked to Kashgar in 2006, the Old City's maze of mud-and-straw buildings was such an exquisitely preserved example of traditional Islamic architecture that it stood in for 1970s Afghanistan in the film *The Kite Runner*. Two years later, in 2008, the crisis of an earthquake in distant Sichuan gave the Chinese government an opportunity to raze Kashgar's historic quarters by claiming them seismically vulnerable. "What country's government would not protect its citizens from the dangers of natural disaster?" reasoned a Han politician, failing to mention that the Old City's demolition was a convenient way of marginalizing the Uyghurs who lived there. Frustrated by a lack of peaceful outlets for anti-government protest, some Uyghurs resorted to bombings and knife attacks on Han Chinese police and citizens, to which the government responded with even more force. Each crisis prompted by "Uyghur separatists" served as an excuse for China to tighten its control over disgruntled minorities, as well as contested frontiers, including by paving border roads so that military convoys could more expediently patrol them. Which is why National Highway 219, the rutted, otherworldly track we'd previously biked across the Tibetan Plateau, leading out of Xinjiang and into the equally oppressed TAR, was closed for construction.

For Ben's sake, I was glad all those roadside mounds of gravel were finally being put to use, but I was also crushed not to return to the Aksai Chin. The nearest alternate route across the Tibetan Plateau was National Highway 109, starting from Golmud, an industrial city two thousand miles away in the Qinghai province. Without enough time on our Chinese visas to bike from Kashgar to Golmud, then continue more than a thousand miles across the plateau to Nepal, we loaded our bikes onto a train. Limited passenger ticket availability meant Mel and I took a different series of trains and buses than the bikes, which we planned to reunite with in Dunhuang, a historic Silk Road trading post. Then we would continue by bus to Golmud. Instead we arrived in Dunhuang to learn the bikes were missing.

A tall, angular Swiss cyclist named Philippe was in the same situation, though he received the bad news with remarkable poise, perhaps because he was a lay monk in the Zen Buddhist tradition.

After meeting in Kashgar and hearing of our plans to bike across the Tibetan Plateau, he'd decided to join us, though he wasn't yet sure how far he'd risk riding toward Lhasa. When Mel and I last snuck across Tibet, in 2006, it might as well have been Shangri-La compared to the North Korea it was now. In the spring of 2008, when Tibetans hoped the world would be scrutinizing China in the lead-up to the summer Olympics in Beijing, protests broke out across the TAR calling for greater independence for Tibet. Chinese security forces responded by firing on unarmed protestors, conducting mass arrests, brutalizing detainees, and torturing suspects—all in the name of preserving national unity. That very year, the government was initiating a grassroots surveillance program called "Benefit the Masses," in which officials were deployed to villages and monasteries throughout the TAR to sniff out dissent among Tibetans and lead propaganda sessions on, among other subjects, "exposing the heinous reactionary crimes of the fourteenth Dalai clique." The official slogan of this Orwellian surveillance initiative, which was originally slated to end three years later but continues still, is "All villages become fortresses, and everyone is a watchman." Everyone but tourists, that is, because China doesn't want outsiders to observe any of this. When we last biked into Tibet, independent travel by foreigners was forbidden on paper; this was now true in practice. Mel and I had no idea whether a stealth crossing of the plateau was even possible, but one thing was certain: we needed our bikes to find out.

As the train company searched for them, Philippe suggested we pass the time in Dunhuang by visiting the Mogao Caves, or the "Thousand Buddha Grottoes," which featured Buddhist frescoes and shrines dating to the fourth century. The tall Swiss cyclist had to fold himself over to fit in the taxi we shared out of the city, but fortunately it was a short ride. The Chinese guide assigned to

us at the grottoes was a pale, slender Chinese woman who introduced herself as Mrs. Chan. She wore sunglasses and carried an
umbrella even inside the dark caves, the better to preserve her
delicate, almost translucent skin through which veins swam blue.
For some reason she was convinced that Mel and I couldn't understand English, though she knew we were from Canada. "Do you
know this term 'flaking'?" she'd say, pointing at peeling paint on
a mural. "It is very technical term, this 'flaking.'" Her own English
vocabulary was full of erratically rich words: she identified some
chubby painted figures in a mural as "auspicious fairies with the
robust bodies." Confusing my suppressed grin for puzzlement, she
said, "Do you know this word *robust*? R-o-b-u-s-t?"

Of the 1,000 original caves honeycombed into conglomerate
rock, only 492 remain. The rest were ruined by earthquakes. Like
buildings at MIT, each grotto is identified by a number, and the
digits marked on the doors collectively give the caves the look of a
trendy apartment complex for the spiritually inclined. Many had
been raided by explorers like Sven Hedin and Aurel Stein, who in
the early twentieth century pilfered priceless manuscripts, hacked
off bits of murals, and stole (or cheaply bought) religious sculptures from the grottoes in the name of archaeological preservation.
These men were knighted for their thievery, which relocated many
of the caves' more portable treasures to museums throughout
Europe, a state of affairs that still has the Chinese justifiably apo-
plectic: the Dunhuang Research Academy's otherwise neutral
book on the grottoes calls Stein and his ilk "despicable treasure
hunters." Fortunately the sheer size of the contents of Cave 96
safeguarded it against plundering: the third-largest stone Buddha
in the world sits inside, tall as a space shuttle. The sculpture gave
off the stillness of a mountain as Chinese tourists elbowed closer
to get a better look.

I thought about how the original Buddha (Sanskrit for "awakened one") never wanted to be gawked at like this. In his lifetime Siddhartha Gautama protested against all forms of iconic representation, worried it would make him seem more divine than human. He wasn't a god but a man, he insisted, and what he taught wasn't a religion but a practical field guide to awakening. In the centuries after his death people respected the Buddha's wishes and depicted him only obliquely, through an empty throne or a footprint on a road. Only starting in the first century c.e. did more anthropomorphic representations come into fashion, the kinds of elaborate frescoes and sculptures that populated these grottoes. "When you see the Buddha, kill the Buddha," said Philippe as we looked at the statue, or tried to look at it through the jostling crowd. "As the Zen saying goes," he quickly clarified, lest we ascribed to him any more literal destructive tendencies.

But where to begin? They were everywhere, whole battalions of them meditating on cave walls, proof that the Silk Road once trafficked in Buddhism as much as trade goods. At first glance, most of the grottoes looked sooty and bleak, yet the beam of Mrs. Chan's flashlight unfailingly revealed them as full of zest and life and colour. I wondered if all darkness concealed a similar complexity, such as the look on Mrs. Chan's face when she pointed her umbrella tip at some bodhisattvas whose faces were gouged out. "By the Muslims," she pronounced soberly, for Islam had supplanted Buddhism along western China's Silk Road by the eleventh century, meaning iconography was condemned—a somewhat perverse reversion to the Buddha's original decree against divine portrayal. Some of the defaced paintings were of holy men, others of wealthy patrons who'd subsidized the shrines' construction, and I found it oddly fitting that no distinctions were made in their destruction, that none were spared for their saintliness or wealth or nobility.

Even more of this troglodytic heritage would've been lost if some wise monks a thousand years ago hadn't sealed up a chamber in one of the caves containing a library of tens of thousands of Buddhist manuscripts, which was only rediscovered in 1900. The grottoes also survived China's Great Proletarian Cultural Revolution, a brand of mass hysteria instigated in 1966 by Chairman Mao Zedong to eradicate the "Four Olds" in China: old customs, old culture, old habits, and old ideas. As many as two million people died across communist China in a decade of chaos and upheaval, during which gangs of students known as Red Guards ran the schools and forced teachers to labour in factories and fields. Education, science, and art were widely condemned as "intellectual" and "bourgeois." Monasteries, temples, books, and works of art were systematically demolished across the country, although Zhou Enlai—the first premier of the People's Republic of China and mastermind behind the road-building project that provoked the Aksai Chin dispute—personally interfered to protect the Mogao Caves from the Red Guards.

Mrs. Chan didn't mention any of this, possibly because she was too young to have experienced the collective insanity that convulsed through her country under Mao, but also, no doubt, because China isn't exactly forthright about acknowledging its past. The Communist Party of China, which still holds power in the country, admitted in a resolution in 1981 that the Cultural Revolution was a "comprehensive, long-drawn-out and grave blunder." But this same resolution goes on to stress that "our achievements in the past thirty-two years are the main thing. It would be a no less serious error to overlook or deny our achievements." Among them, of course, was the "peaceful liberation" of Tibet.

Mrs. Chan snapped her umbrella closed, as though the storm she'd been waiting for had passed. The tour was over. We shuffled into the bright heat and Philippe folded himself into another taxi. As we drove back to Dunhuang I couldn't stop thinking about the caves, which seemed seared on my inner eyelids, for I saw a thousand buddhas every time I blinked.

At the guest house a heartening message awaited: the train company had located our bikes in Lanzhou, in China's far east, and they were being shipped directly to Golmud. Mel, Philippe, and I packed our gear and took an overnight bus to meet them there.

The bus lurched past what smelled like a chemical factory and looked like a mine, the earth seeping with bright liquid sores. Even after dusk the heat was vicious, and I envied the man a few bunks ahead of me for being able to take off his shirt. His back was as pale and smooth as a teenager's, but when he turned around I was shocked to see a face decades older, a rough pumice of wrinkles and scars. To my right a portly Chinese man cuddled with a woman half his girth and age. "Ah, young love," Mel whispered when I pointed them out. Blame the chemicals in the air, or sleep deprivation from a bus whose coffin-size bunks and rollicking motion seemed engineered to keep passengers awake, but in any case, when we got off in Golmud, I left Mel's helmet on the bus and only realized it after the bus drove away.

Pedalling the high passes of Tibet without a helmet wasn't an option, so our first task in Golmud was searching for a replacement. Unfortunately, the only options we found were for motorcycles—hulking, full-face carapaces that weighed at least ten pounds and came with chin guards and tinted visors. Mel gamely agreed to wear one of these on the road to Lhasa, declining my offer to let her wear my own sweaty, filthy bike helmet. "This one is very robust," she remarked from inside a cavernous motorcycle helmet. "Do you know this word? R-o-b-u-s-t?"

A kind young Chinese man in the shop gleaned that our two-wheelers weren't motorized and escorted us, after several detours and dead ends, to a bona fide bicycle shop. Mel tried on an aerodynamic royal blue helmet and I wandered around the shop, admiring the crisp newness of everything: inner tubes, spare wheels, at least seven different styles of bike shorts. Then I saw a bulletin board in the corner that was covered in photos, several of which depicted cyclists fist-punching the air in front of the Potala Palace. "It's very popular," the shop owner remarked when he noticed what I was looking at. "Many, many Chinese people bike the Qinghai-Tibet highway." The bikes in the photos flew bright red Chinese flags, and the cyclists wore helmets, dark glasses, and polyester face-masks, presumably for sun or dust protection. I couldn't tell at a glance whether any given cyclist was female or male, young or old, Canadian or Chinese . . .

"Do you sell face masks and flags like these?" I asked, trying to keep the excitement out of my voice.

I could hardly breathe through the polyester as Mel, Philippe, and I biked out of Golmud that evening, but I looked so anonymously Asian I didn't mind. The road was quiet, with only the occasional transport truck to send our Chinese flags flapping. Sedimentary mountains tilted up from otherwise flat terrain. The late afternoon light was so intense it seemed like a hard fact in the air, the sort of thing I could hold on to and steady myself by. I'd need that kind of support with the first checkpoint of dozens on the way to Tibet just ahead, though we didn't really have to worry, given we weren't in the TAR yet.

Sure enough, the police paid us no mind as we biked under the raised guardrail. As we accelerated away I switched to a higher gear and my chain jammed behind the chain ring. I stopped just beyond the checkpoint to fix it, with Mel and Philippe keeping watch, but no amount of yanking would release the wedged links. I finally had to break them, remove the rear wheel, and wrench off the gears to get the chain free. Then I fixed it with a replacement link. In the process I bent a spoke and threw my rear wheel out of alignment, though I didn't notice until the next day, when I could barely keep up with Mel and Philippe. I couldn't figure out why: Was it the steady gain in altitude? Biking almost fifty miles after weeks off? Only when we stopped to camp and I wheeled my bike off the road did I realize the brakes were rubbing on my wheel.

We pitched our tent beside a river, under a bridge that carried the high-speed train to Lhasa. Built over permafrost at high altitudes, this railway was as controversial as it was complicated to engineer. A Chinese pop song lauds the Qinghai-Tibet railway as "an amazing road to heaven, carrying us to paradise," and it seems clear to whom the "us" is referring: ethnic Han Chinese workers, who took the railway to China's "wild west" in droves following its completion in 2006, depriving Tibetans of jobs and opportunities and further consolidating Chinese control over the TAR. Staring up at the tracks, I saw someone standing on them, waving his arms. I waved back. Shortly after, a dozen men in military camouflage showed up.

The Chinese soldiers hovered around us, inspecting our tents and bikes, and asked us questions in Mandarin we conveniently couldn't understand. When one of them made a call on a hand-held radio, it seemed obvious that trouble was on the way. Despite the sick feeling in my stomach I went on fixing my wheel. Half an hour later another military man in uniform showed up, and judging from his fancier camouflage and decorated lapels, he was some kind of officer or commander.

"Hello! Welcome to China!" he said in flawless English. He proceeded to explain, very apologetically, that we were camped on military grounds that were somehow related, from what I could gather, to the train tracks. "I'm afraid we must request that you move at least a half a mile up the road."

Not a problem, we reassured him, feeling intensely relieved.

"I wish you all a good journey," he continued warmly. "Is there anything I can do for you?"

I considered asking for Alien Travel Permits but decided that would be unwise. For all he knew, we planned to turn off the Qinghai-Tibet highway long before the TAR. So far, our ride was perfectly legal.

After the soldiers left, we biked farther up the road and pitched our tents at a Buddhist temple with the permission of the care-taker, an elfin Chinese man no taller than my shoulders. His smile matched the upswing curves of the pagoda when Philippe got out his monk's rakusu, a vest-like garment indicating his lay ordination in the Zen Buddhist tradition. Across the valley a giant yellow bull-dozer energetically stripped away the slopes of a mountain, work that I assumed would stop by dusk, but as darkness fell the spot-lights came on. When I peeked out of the tent that night I saw freshly scraped mountain slopes glowing white as teeth, and the sputtering and drilling of heavy machinery continued through dawn. I woke up with a headache and blamed the altitude.

In truth we weren't very high yet. Only a couple days later did we crest the first of a dozen passes across the Tibetan Plateau, the road rising through mountains whose dirty glaciers looked like ice sifted with cinnamon. I spent most of the climb fantasizing about freshly baked cinnamon buns. At the top of the pass Mel dug through her panniers and pulled out a box of stale chocolate-filled buns instead. But when we bit into them, there was no chocolate, just a hole where the chocolate should've been. It was almost like a homecoming.

"The number one ingredient in packaged food in China?" said Mel. "LIES."

"The number two?" I said between bites of unbearable sweetness. "SUGAR."

Given that the Chinese energy bar we ate next listed "loosening" and "meat flavouring" on the label, we sure hoped lies and sugar were the primary ingredients, though the latter would mean I'd finish this Silk Road with nine cavities for souvenirs. We offered some snacks to Philippe, who politely declined, preferring instead to preserve his smile. By now he'd decided not to risk the ride to Lhasa. Instead, he would head toward Yushu, a part of the plateau that was ethnically Tibetan but beyond the TAR, meaning he didn't need a permit and guide or stealth tactics to travel there.

Mel and I were sorry to see the Swiss lay monk go, especially when the quiet gravel side road he turned down looked far more appealing than the paved highway we faced, edged by enormous powerlines that made the sound of knuckles cracking. To reach Nepal before our Chinese tourist visas expired, Mel and I needed to cover fifty miles a day on this road, across relentlessly high altitudes, into fierce headwinds, wearing face masks, dodging the law, for twenty-one more days. No time off, no room for error, and no resupplies—not because grocery stores and restaurants weren't widely available in eastern Tibet, but because we didn't dare stop for fear of getting caught. For the same reason we concealed our camp each night, which wasn't easy in a land of huge horizons.

At least biking in stealth mode became less suffocating once we cut breathing holes into our face masks. Whether we smiled or rode along with our mouths open, breathing hard, the mask gave us each a leering rictus, which seemed an appropriate greeting for Chinese police cruisers. At first I panicked at the sight of those gleaming white SUVs with red lights on the roofs, but none slowed to stare as they passed: we blended in with the sundry Chinese cyclists on this road, though to my eyes our distinctions were obvious. For one, their bikes usually had only two panniers, compared to our four, because they were able to eat at restaurants and sleep at guest houses along the way. They also tended to pedal in an extremely high gear at a hard grind, compared to our speedy twirling in a lower gear. And despite this being a paved highway, the Chinese cyclists all rode mountain bikes with at least front, if not dual, suspension. The sole exception we saw was a forty-something mother from Beijing, who rode a folding bike with a few piddly gears. Somehow she managed to keep pace with her teenage son, who rode a mountain bike, at least for the hour or so Mel and I rode alongside them at their invitation. They'd wanted to practise their English, which they knew we spoke from the moment we said "*Ni hao!*" to them with an obvious accent. But if these Chinese cyclists were aware that we shouldn't have been on that road without a guide, they kindly didn't say so.

Despite our resolution to not stop anywhere, all vows have a breaking point. Mel's and mine was a Tibetan antelope rehabilitation centre on the side of the highway. The reward for our flagrant incaution was being mobbed by dozens of orphaned baby antelopes with bulging brown eyes and adorably matted tan fur. Their stilt-legs always looked bent, so that they seemed to walk on tippy toes, and they nibbled delicately on grass as if they weren't sure they liked the taste. On our previous bike trip, we'd seen wild Tibetan antelope, or *chiru*, waft like smoke across the western plateau. This migratory species is endemic to the Tibetan Plateau and severely endangered from poaching for their soft, silky underwool, or shahtoosh, which ounce per ounce is worth more than gold and diamonds. To their credit, the Chinese government has cracked down on illegal poaching, and they'd generously raised sections of the Qinghai-Tibet railway to allow passage to the nomadic herds. Unfortunately, the state had also embraced wilderness conservation as a convenient excuse to force Tibetan nomads off traditional lands and into soulless subdivisions.

Another day, another pass, each higher than the last. Some days we did two passes, both over 16,400 feet, such as when we formally left Qinghai and entered the TAR. The border is marked by Tanggula Pass, which evocatively means "mountain on the plateau," but in reality was a desolate circus of cars and trash. The plastic wrappers and bags fluttering on the pass were parodies of prayer flags. A stone monument depicted what looked like a pair of soldiers, one of them talking on a telephone. Piles of human excrement sat at the base of it. Mel and I left as quickly as we could.

It wasn't until a few days later that we felt more truly back in Tibet, when we crested the top of another pass that was marked with prayer flags. Yak skulls carved with *"Om mani padme hum"* were tangled among them, and there was no trash, no crowds, no people until a few Tibetan men showed up on motorcycles, looking like funky cowboys in their wide-brim hats, dinner jackets, and pointed leather boots. They turned coldly away, and only as they drove off did I suspect why. Of course. I looked Chinese.

The flags we flew and the face masks we wore were essential, the only way to bear witness to the Tibet that China didn't want us to see: the checkpoints to restrict Tibetan movements at the entry and exit of each town; the traditional homes flying Chinese flags, bright as bloodstains against the white stucco buildings; the concrete statues of police officers looming at regular intervals along the road, their faces frequently smashed featureless, their necks slung with beer bottles and white prayer scarves; and the road signs featuring the Tibetan language in tiny script below much larger Chinese characters, reinforcing subordination right down to font size. But as we rode through towns we were too paranoid to stop in, camped out of sight each night, and met Tibetans on high passes who didn't even glance our way, I felt like I was living at a subtle yet ineluctable remove from "reality." What I saw in Tibet was not Tibet but some blurred, partial rendition of the place, distorted by the limits of my mode of exploration. I was effectively behind Plexiglas.

Alexandra David Néel also disguised herself to see this forbidden land, but it dawned on me that she was a Buddhist, meaning her pilgrim's garb reflected a deeper truth. Repeating *"Om mani padme hum"* across the plateau was genuine prayer. As we biked past Tibetan families gathering barley from stubbled golden fields, or flossing mountain passes with prayer flags, the urge to explain myself, to apologize, fevered through me. Instead I kept quiet and kept pedalling, and in the silence of my passing I could hear the pennant flapping on the tail-end of my bike, a kind of shame trailing red and loud behind me.

Roughly halfway through Darwin's journey on the *Beagle* he was ready to quit. The seasickness, isolation, and physical strain were grinding him down. "I am sometimes afraid," he confessed in a letter home, "that I shall never be able to hold out for the whole voyage." I was sometimes afraid of the same thing halfway across Tibet, where all the wrongness in the world seemed wide-angled. Holing up in a countryside cottage began to sound awfully appealing, especially when the imperative to hide our tent meant we camped, at one point, in a garbage pit. It was full of random debris—broken crockery, old pill packages, a plaid shirt, lonely shoes. That night a storm flung rain at the tent with such force it sounded like bits of gravel. At any moment I expected to be buried alive, and I was too tired to care. After just a few weeks on the Tibetan Plateau, Mel and I looked gaunt, strung-out, with leg muscles like wads of gum stuck on bone.

"Every body part aches," Mel groaned the next morning on a rest break. "But I've learned the secret to overcoming this . . ."

"Do tell," I said.

"Don't listen to your body!"

But when my mind screamed *Stop* now, my legs chorused their agreement. I missed jeans, couches, pizza. I missed waking up eager to see what lay around the next bend, rather than doubting whether I wanted to know. I missed feeling like a real person, rather than a ghost, something I often complained about to Mel. But at my first opportunity to interact with Tibetans I promptly melted down. One day we couldn't find anywhere flat or hidden enough to camp, so we hesitantly accepted a family's invitation to pitch our tent behind their home. Mel and I worried about implicating them in our illegal stunt, but the campsite was hidden from the road, among yaks grazing in shaggy skirts, so it seemed unlikely we'd draw attention. It was a glinting, gorgeous day, and barely mid-afternoon when we set up the tent, but I crawled into my sleeping bag and couldn't bear getting out again. I felt evicted from the trip by tiredness, as despairing and world-weary as Rimbaud— enough seen, enough known, enough had, *enough*.

But even lying there with my eyes closed I heard the Chinese flag flapping from the Tibetan home, not to mention the ones on our own bicycles, sprawled in the grass outside. I could picture military convoys trawling the highway and spewing propaganda. I imagined nomads being herded into subdivisions while chiru ran free in wilderness reserves. I saw a chilling, sterilized land in which no horizon is unmarred by authority, no movement goes unmonitored, and every hint of peaceful protest is crushed beneath the heel of the state. And there I was, joyriding through this oppressive landscape, a tourist in a regime Tibetans literally set themselves on fire to protest and escape. Eleven people self-immolated the year of our bike crossing—men and women in their twenties or thirties who shouted, "Tibetan people want freedom!" or "Long live the Dalai Lama!" before pouring kerosene over themselves and lighting a match. Eighty-six did the same the following year, which was also when the Chinese government confiscated passports from TAR residents and made it extremely difficult for them to obtain new ones, effectively imprisoning six million Tibetans. Such injustices sickened me, as did my relief at being able to leave. I couldn't wait to leave. Mel went to visit with the Tibetan family without me, explaining that I wasn't well.

Yaks grunted and snorted around the tent, munching closer and closer. I shook the tarp to provoke a retreating thunder of hoofs. Against the distant drone of traffic I could hear the delicate pinging of flies trapped between the tent's inner and outer walls. I lay in my sleeping bag, aching all over, and fervently hoped humans never made it to Mars. We didn't deserve a new world; we'd just wreck it all over again. As a kid I'd genuinely believed that the discovery of alien life, whether sentient beings or microbes, would change lives, incite a revolution near-holy in its repercussions. At the very least people would be kinder to each other, knowing we're all of a kind, earthlings every one, whether Turkish or Armenian, Indian or Pakistani, Tibetan or Uyghur or Han Chinese. We'd collectively awaken to the fact that we're all lost in this mystery together.

Now I wasn't convinced. Discovering extraterrestrial life wouldn't change a thing, just as learning to fly didn't lift us higher as people, just as Voyager's pale blue dot photograph failed to dissolve nationalism the way it should have if we'd *truly* seen it. "Look again at that dot," Carl Sagan pleaded. "That's here. That's home. That's us. On it everyone you love, everyone you know, everyone you ever heard of, every human being who ever was, lived out their lives . . . on a mote of dust suspended in a sunbeam." Meanwhile we've discovered microbes eating sulphur in boiling vents at the bottom of the ocean, Earth-size exoplanets orbiting distant suns, proof everywhere of the rarity, ingenuity, and resplendence of life in the universe—and such facts haven't budged our priorities an inch. What is the point of science and exploration if people persist in living and dying as they always have, namely selfishly, obliviously?

Maybe infinity begins at the point we can't see past, can't love past. How small we are when this point is ourselves. The problem with borders, I was beginning to realize, isn't that they are monstrous, offensive, and unnatural constructions. The problem with borders is the same as the problem with evil that Hannah Arendt identified: their banality. We subconsciously accept them as part of the landscape—at least those of us privileged by them, granted meaningful passports—because they articulate our deepest, least exalted desires, for prestige and permanence, order and security, always at the cost of someone or something else. Borders reinforce the idea of the alien, the Other, stories separate and distinct from ourselves. But would such fictions continue to stand if most of us didn't agree with them, or at least quietly benefit from the inequalities they bolster? The barbed wire begins here, inside us, cutting through our very core.

I heard footsteps approaching the tent. The door zipped down and Mel poked her head in.

"You okay?" she inquired. "I thought I heard crying."

"Must've been the yaks," I lied, thinking, *All explorers must die of heartbreak.*

The oracle-bone script of ancient China recorded the world as it plainly was: a fire was a fire, a fish a fish, a mountain a mountain. This earliest form of Chinese writing was so pictographic that you can almost guess at the meanings of divinations carved into turtle shells and animal bones during the Shang dynasty of 2000 BCE. Modern Chinese characters evolved from these symbols, and though the current incarnation of that writing is less literal, more esoteric, you can still read a lot into it. The symbol for *gradually*, for example, is based on pictographic forms that showed water cutting through stone.

This has been the Dalai Lama's gentle but persistent approach to freedom from China. Although the Chinese government vilifies the exiled Buddhist leader as a terrorist, the "simple monk" I listened to at Oxford wasn't even advocating for Tibetan independence, merely a truer autonomy for Tibet within China. His non-nationalistic, non-violent approach frustrates certain Tibetans, who see it as too soft, a cop-out, when what they want is their country back. The Dalai Lama has now "retired" as the head of the Tibetan government-in-exile, and he instituted a democratic system of governance in his place, meaning his leadership responsibilities now are strictly spiritual. And in this capacity, as the human incarnation of the bodhisattva of compassion, his job is to end the suffering of not just his people, but *all* people. Instead of seeing good Tibetans and bad Chinese, as Pico Iyer noted in *The Open Road*, the Dalai Lama sees *potentially* good Tibetans and *potentially* good Chinese. I tried to be that nuanced and open as I biked across the Tibetan Plateau, but everything I saw built a forceful case for bias.

It was only late August, but the poplars were already flaring gold. Fallen leaves crunched beneath our wheels, and the paper prayer flags scattered on mountain passes made a similar noise when we biked over them. Tibetans threw the colourful squares into the sky in a bid for good fortune, and if nothing else, this had the immediate effect of collaging a dark road into something brighter. On one pass a bus drove past me just as its passengers threw the papers out the window, so that prayers stormed down all around me. One of them caught on the brim of my helmet without ever hitting the ground. I tucked the gritty, sage-coloured square into my journal for good luck. Depicted on it was a wind horse, or *lung ta*, a pre-Buddhist symbol for inner wind or positive energy shown as a horse lugging a jewel on its back. When someone's *lung ta* decreases, the Tibetans say, they are grounded by negativity, and when *lung ta* increases they see things more positively and soar. "The very same thought can lead to a state of freedom or to a state of confusion," wrote a Tibetan monk, "and the direction it takes depends upon *lung ta*."

Headwinds made descending the pass almost as hard as climbing it. I willed them to change direction, and when that didn't work, tried nudging my wind horse another way: Why should this bike trip be easy on me? I'd read about how the Chinese character for *presence* is derived from earlier pictographs showing a hand eclipsing the moon, which made me wonder if I hadn't been giving Neil Armstrong enough credit. Maybe what the astronaut had really been saying, in covering the earth with his thumb during the Apollo landing, was something like, *Be present, utterly present. This world deserves your deepest attention.* Read along similar lines, what a difficult road says is, *Wake up. Keep your eyes focused on what's bigger than the sadness directly in front of you, the sadness that both hides and gestures toward some larger enigma.* So I did, which is how I saw the pilgrims.

You couldn't really miss them, flattened on the pavement as though hit by a car, except for the relieving and life-affirming fact that they periodically stood up. The Tibetan man and woman placed their palms together at their chests, raised them to their crowns, and lowered them to their foreheads, throats, and hearts in a fluid sequence of gestures. Bending at the waist, they slid their hands, knees, bodies, and then foreheads to the pavement, with traffic screeching a few feet away. Then they stood up, took a few steps forward, lifted clasped palms above their heads, and repeated the ritual. They would repeat it all the way to Lhasa or enlightenment.

Mel pulled off her face mask and sunglasses when we caught up to them and I did the same. Their eyes widened at her freckles and my dirty blond hair, then they laughed. We shook hands, beamed at each other, and exchanged what few words we had in common. Mel and I prided ourselves on travelling light, with just enough warm layers, camping gear, and instant noodles to last us across the Tibetan Plateau, but these two carried nothing but the clothes they wore. Thick wool sleeves protected their arms, leather aprons their knees, and wooden paddles their hands. We offered them a Snickers bar and I tried not to stare: in the middle of each of their foreheads was a coin-sized callous, a third and unblinking eye, caused by the friction between pavement and bone.

Eventually we said goodbye, *tashi deleg*, and continued down our respective roads. In my handlebar mirror I watched them shrink on the highway behind us. I held my breath as transport trucks swerved near their prone figures, and whistled with relief when I saw them rise again. Dust and fumes rose like incense from the road. Speeding vehicles sprayed pebbles into the gutter. With every step, every repetition, the calluses on their brows must have grown thicker, denser, the skin hardening to a darker and more permanent shine. Sometimes scars are a kind of protection, making prayer possible. Sometimes even wilderness needs a wall. The pilgrims disappeared from view and I pedalled on, nothing in my pockets but stories, wind, all kinds of weather.

As we biked toward Lhasa, the most policed stretch of the Silk Road yet, convoys of army trucks fumed past spewing exhaust from tailpipes and propaganda from loudspeakers. I held back the urge to yell, *Shut up*, another expression included in the historic Silk Road phrase book, evidently as relevant in twelfth-century Tibet as it was today. We raced downhill through a gauntlet of checkpoints, swerving around the vehicles forced to stop at each one. After what felt like hours we sped into the sacred heart of Lhasa, where "the inhabitants of the city all adopted foreign dress, and submitted to the enemy; but each year when they worshipped their ancestors, they put on their clothes, and wept bitterly as they put them away." Except that this statement, which seems to portray Tibet's capital city under Chinese rule, actually refers to Tibetan-occupied China.

In the seventh and eighth centuries, the Tibetan Empire conquered a number of Chinese outposts in its quest for territorial expansion. Among them were Dunhuang, and it was that city's colonized Chinese residents, not Tibetans in modern Lhasa, who are described in the passage from the royal annals of the Tang dynasty. Other documents recovered from caves in Dunhuang reveal the tensions between the Chinese and their Tibetan overlords. In one government letter, a Tibetan minister addresses petitions against the habit of Tibetan officials kidnapping Chinese women to be their wives. "To his credit," notes Sam van Schaik, a scholar of Tibetan history, "he responded by banning the practice of kidnapping, saying that the women should be able to marry according to their own wishes." In another exchange, after a Chinese uprising against their Tibetan masters, a Tibetan minister curtly dismissed requests by Chinese officials for greater powers, and instead outlined the strict hierarchy of positions within the government. "The long list is a treasure-trove for those who study the bureaucracy of the Tibetan empire," writes van Schaik. "But let us just note one thing: the letter makes it clear that even the lowest-ranking Tibetan is of higher status than the highest-ranking Chinese."

Tibet's own history, then, is blighted by acts of greed and colonialism. I knew taking sides in any modern geopolitical conflict involves, by default, some degree of historical amnesia. This is for better and for worse: forgiving means forgetting, but also, sometimes, forsaking inconvenient facts, such as the Sino-Tibetan peace treaty signed in the ninth century. In it the king of Tibet and the Tang emperor agreed that "both Tibet and China shall keep the country and frontiers of which they are now in possession," and that "from either side of that frontier there shall be no warfare, no hostile invasions, and no seizure of territory." The treaty was inscribed in both Tibetan and Chinese script on a stone pillar near the Jokhang Temple in Lhasa, an ancient pilgrimage site, "so that it may be celebrated in every age and every generation."

Mel and I didn't pause in Lhasa long enough to take in the pillar, though reportedly it still stands, its treaty text a little more weathered and unreadable each year in a sad marriage of political and geological erosion. A more recent monument commemorates the "Peaceful Liberation of Tibet," but we didn't stop to see that either. Designed to look like a simplified, concretized Mount

Everest, the statue, from the photos I'd seen, bore as much resemblance to Chomolungma—the Tibetan name for the mountain, meaning "Mother goddess of the world"—as the bluntness of a Boeing 747 does to the grace of a bird. Unsurprisingly, the monument's inscription fails to mention that liberating Tibet involved military air strikes on monasteries. "The Tibetans saw giant 'birds' approach and drop some strange objects," reports Chinese-born writer Jianglin Li, "but they had no word for airplane, or for bomb." (They do now, at least for airplane: *namdu* means "sky boat.") Mel and I didn't stop anywhere in Lhasa. Our only goal was getting out again.

Alexandra David-Néel didn't linger much longer, relatively speaking. After spending half a lifetime trying to reach Lhasa, which she did in February 1924, her beggar's disguise prevented her from moving in the intellectual and spiritual circles that most fascinated her. She and her adopted son, Yongden, left just two months later, but she spent the rest of her life making sense of what she'd seen on the plateau—or not sense, exactly, so much as an evocative written record of Tibet's magic and mystery, invaluable testimony to what existed on the plateau before the Chinese took over.

I recited her name in a kind of mantra as I followed Mel out of the city, past tourists and police cars, past strip malls and bars and discos, past neon signs insisting on the brightness of everything while advertising brands popular in Beijing. At one point I heard a yell behind me and pedalled harder, imagining myself arrested and forced to confess everything. Yes, officer, I have set off for distant worlds without the means or intention to return. No, officer, I have not taken a single breath of this life or any other for granted.

But nobody noticed as we fled the fabled city. For a while the road followed the Kyi River, a tributary of the Yarlung Tsangpo, the longest river in Tibet and the upper section of the Brahmaputra.

After a few hours we saw a police officer statue on the side of the road ahead, his face expressionless, his one concrete arm raised in a stiff command to stop. As Mel rode by she gave him a mock high-five.

The river and road squeezed through a narrow gorge over the next few days. Traffic was minimal except for livestock. At one point a yak leapt out of a ditch directly in front of me, so I slammed on my brakes as did the bus approaching in the opposite lane. The yak was fine, but my front tire hissed flat. I stopped to fix it while Mel kept an eye on the creature, now grazing placidly in the ditch on the far side, sweeping the grass with the black broom of its tail. There was a thorn in my front tire, so I pulled it out, patched the inner tube, and pumped it up, only to discover another thorn in my rear tire. When I tried to extract it, the tube deflated. As I patched this second flat, I saw that the sidewall of my rear tire was ripped and bulging, distending the wheel out of true. This was my "new" tire, lugged all the way from Istanbul and swapped into rotation in Tajikistan when my previous tire wore thin. We had no more spares.

A Chinese cyclist, the first we'd seen since before Lhasa, caught up with us and shared some Tibetan bread slathered with hot sauce. He was also heading to Nepal, he told us in broken English, and we knew he'd beat us there when we heard how far he biked each day. "Sixty, maybe ninety miles?" Mel and I stared at him: we were barely managing fifty. The three of us rode together for a few hours, which is to say he left us behind on the climbs on his lightly loaded mountain bike, and we passed him on the descents thanks to the heavy ballast of our expedition gear. When Mel and I decided to camp at two in the afternoon (we'd been biking since before dawn), he seemed mystified. "But what will you do now?" he asked, as if there was nothing to do on a bike trip besides bike. Read, write, nap, we told him. Watch the Tsangpo surge by the tent, that torrent of silt and sound. Rest up so we can wake up and do it all over again tomorrow.

The narrow river canyon widened into a broad valley with a Siberian flair: pale blue skies, spiky clusters of trees, a wide, sluggish churn of water seeking the sea. In some places the river flooded its banks, swamping trees up to their lowest branches, so that they looked like people with their arms raised in protest or surrender. Signs of industrial and military activity were rampant: train tracks under construction, some kind of immaculate air force or military base, mysterious explosions across the valley when we camped at night.

Yet there were fewer Chinese flags on display southeast of Lhasa, especially past the second-largest city of Shigatse, as well as fewer police cruisers. Tibetan villages wafted the incense of burning juniper, sage, and yak dung. Neat fields of wheat and barley flourished in every flat place, and women sang in high warbling voices as they worked in them. Some children ran beside us for the final half-mile up a high pass, barely panting as we breathed hard in the thin air. At the top we saw sharp, glinting peaks in the distance, but clouds masked Chomolungma. The road down twisted and turned beside a turquoise river, then stopped just short of Shelkar, or New Tingri, where we waited until dark to sneak across the most daunting checkpoint of all.

According to what we'd gleaned from cycling forums and blogs, all vehicles, including bikes, were forced to stop at this checkpoint to show papers. An Australian cyclist had biked this route with a guide as part of a tour group, looking for loopholes to avoid having one in the future. He concluded that this checkpoint was impossible to sneak across: the military complex had two guardrails and immediately following it was a bridge over a river, so that to sneak around required fording that flow. Earlier that spring, an American cyclist successfully snuck from Golmud to here without a guide or a permit, but he was so daunted by this checkpoint that he gave up and turned himself in. In his case, the guards glanced over his passport, detained him for a while, then let him go, possibly because he was two days from exiting Tibet anyway. But after coming so far, Mel and I didn't want to risk being stopped now, plus the American had mentioned a promising lead: apparently a dirt track veered off the road just before the checkpoint, possibly offering a detour. The problem was we couldn't find it. Everything else about Shelkar matched the cyclist's vague description of the checkpoint: It was just after a rise, in the main town before the turnoff to Everest Base Camp. There was a guardrail, a long corridor of buildings, and a bridge over a river on the far side of town. All that was missing was the dirt track.

To the left of our hiding spot a flock of sheep and goats with bells around their necks grazed, sounding like wind chimes. A scrawny Tibetan shepherd in his teens wandered over to the gravel ridge we crouched behind, trying to stay out of sight until nightfall. The boy sat on his haunches about ten feet away, staring at us expressionlessly. If we hadn't been so exhausted, terrified of the night to come, and worried he might alert someone to our presence, Mel and I would've been friendlier. As it was we ignored him, which was easy because he made no attempt to interact with us. He just sat and stared as we wrote in our journals, then dozed, then cooked the usual round of instant noodles for dinner. It was a relief when he finally walked away to tend to his sheep.

Later that evening, though, I ran into him while scoping out a possible "high route" above town, one I hoped might circumvent the checkpoint by hugging the grassy slope to the left of the main road. He waved me over with a friendly grin and offered me his soda bottle, which was full of a white, yogurt-like drink that tasted refreshing and slightly bubbly. He insisted that I keep the bottle and give the rest to Mel, who was back with the bikes, then waved goodbye and hustled over to his flock. I walked back to our hiding spot sickened by the fact that we'd eaten right in front of him and hadn't offered to share, as if we couldn't spare the food or the time, as if we'd been wearing face masks for so long they stuck to us even after we'd taken them off.

There was no point in wearing face masks or flying the Chinese flags that night. Darkness was the only disguise we needed. It was drizzling when we set off at midnight. Since there was no road or trail we had to heave the loaded bikes across the uneven slope, occasionally flicking our headlights on and off again to glean the lay of the land without exposing ourselves. We stumbled blindly for an hour until we reached the far edge of town, long past the checkpoint, and crossed the bridge leading away. Home free.

A transport truck roared past us, its high beams revealing the twists and turns of the road, which angled gently downhill so that we surged along it without pedalling, without any sense of distance or dimension. So many stretches of the Silk Road were more familiar to me as constellations than countries. When the truck came to a stop a mile or so ahead, where some lights flickered, we figured that had to be the turnoff to Everest Base Camp. As we got closer, the lights resolved into a corridor of buildings blocked by guardrails: the checkpoint we thought we'd already circumvented.

Mel and I quietly chorused a range of curse words, then backtracked to search for the rumoured track before the checkpoint. Sure enough, as the American cyclist described, a dirt track angled off from the main road a quarter of a mile before the buildings. We rode down it until the terrain got so muddy we had to walk the bikes.

This turned out to be a good thing, because it meant less weight on the tires when we rolled over the broken glass that suddenly glittered everywhere, a spill of stars. For the record, for the sake of all future explorers, let it be known that this dirt track dead-ends at a dump.

We struck off cross-country toward the river, collecting pounds of mud on our shoes and wheels in the process. As we travelled the dark side of the Silk Road, I had the alarming sensation of bumping up against something huge and shapeless, over and over again. Perhaps it was just my heart trying to batter its way out. We reached the shore only to realize the river was too fast and deep to ford. The only way across was the bridge immediately after the checkpoint, which we could see now by the light of a guardhouse. That same building cast the only shadow available for cover, so we crept into it, unnervingly close to the soldiers I could see milling around inside. I hoped all they could see, if they glanced out the window, was their own reflection.

Mel and I waited at the bottom of the embankment leading up to the bridge. It was maybe ten feet high with what seemed to be a sixty-degree slope, though these dimensions were no doubt warped by nerves. A truck passed, and in the cover of its noise Mel climbed up to the road, then disappeared into darkness across the bridge. I tried to follow her up the embankment but my bike was so freighted with mud I could barely move it. I got halfway up before sinking to my knees, fully exposed in the light of the guardhouse. Then a vehicle started up, and fear volted through my veins. I heaved the bike's front wheel onto the road, squeezed the brakes to hold ground, and hauled myself up. Only when I'd pushed beyond the glow of the guardhouse did I stop to catch my breath, which is when I saw a wedge of light widening behind me.

I jumped on and tried to pedal, but so much mud embalmed

my chain and derailleur that the cranks wouldn't turn. I clawed at the mud and tried again. The bike lurched forward, gaining speed as it shed weight. There was no sound except for mud hitting the guardrails of the bridge, my heart, some dogs barking, the *swish, swish* of my splash pants. I accelerated past Mel—who was waiting in the darkness on the far side of the bridge—and into a ditch. She followed a few seconds later, just before the vehicle passed. We looked up to see a dozen Tibetan men and women sitting on piles of stuff in the rear of a pickup truck, staring down at us in astonishment. Every heartbeat is a history of decisions, of certain roads taken and others forsaken until you end up exactly where you are. I raised my hand to wave but the vehicle roared off into the night.

Nothing remained between us and Nepal except two breathlessly high passes followed by a two-vertical-mile plunge off the Tibetan Plateau, down through the clouds. We arrived in Zhangmu a few days later, just after the Chinese-Nepal frontier compound had closed for the day. It was pouring rain. The warm air smelled of soaked earth and flowers and green growing things, almost honeyed in its richness, its surfeit of oxygen. We stood at the closed gate of the Chinese border compound, soaked and exhausted, until the guards took pity on us and invited us to stand under the roof.

"Passports?" one of them asked, so we handed them over.

"Where is your Chinese guide?" another inquired. Mel dug around in her backpack and pulled out a guidebook, beaming a bland innocence.

The guards were not amused, but they were eager to go home. They ushered us into a small office where one of them gestured that Mel and I should sit down on a pair of rather posh leather chairs. We declined, pointing out how wet and muddy our splash pants were: we'd ruin the leather. The officer smiled for the first time and then firmly reiterated that we were to sit down. Once we did, he typed details from our passports into a computer in the office. After a while he disappeared with our documents, closing the door behind him.

Mel and I waited in the tiny room without speaking. The clamminess of my clothes felt wetter if I fidgeted so I stayed still, though my mind was bouncing off the walls. The most severe consequences we'd heard of cyclists facing after sneaking into Tibet was getting booted out again, but in recent years, few people had managed to travel here independently at all, and Chinese policies changed all the time. There was no guarantee they wouldn't make an example of us. If it came to that, I vowed to write about Tibet from behind bars if necessary, whatever it took to fulfill my second duty as an explorer. Even Marco Polo had managed to dictate his travel book from jail, and lacking a laptop or paper, I could scratch oracle-bone-like symbols into the walls, conveying thousands of kilometres of the Silk Road with a few symbols: wings, fences, stars, a mote of dust suspended in a sunbeam.

The guard opened the door and handed us our passports. We stared uncomprehendingly. "You can go," he clarified in a voice so sincere and large-hearted I was sure he was lying. Mel and I thanked him, just in case he wasn't, and did our best to wipe the leather seats clean with gloves that were almost as muddy as our pants. Then we scooted outside as fast as we could without looking like we were trying to get away with something.

I kept expecting to be called back, to be scolded, to at least pay a fine, but nobody bothered us as we left the Chinese border compound and wheeled our bikes very calmly toward Friendship Bridge, which links Nepal with Tibet across the Bhote Koshi river. The air was dense and humid, a hot, wet felt against my face. The metal cleats on my bike shoes clacked dully on the damp concrete of the bridge. I thought about David-Néel, tried to mimic the wilderness of sky in her step. I thought about the pilgrims, moving through the world as if their most immense responsibility to it was wonder. The bridge seemed to go on forever, though it was only a hundred and fifty feet. On the far side Mel and I stopped at the first restaurant we saw and ordered everything deep-fried on the menu.

11.

ROAD'S END

Indo-Gangetic Plain and Greater Himalaya

The Bhote Koshi melts from the glaciers of Tibet and flows into Nepal, just west of Chomolungma, twisting through steep gorges into which villages are hammered like tiny wedges of humanity, trying to pry the Himalaya apart. The river joins six others in Nepal to become the Sapta Koshi, which eventually meets the holy Ganges in India, and merges with the Brahmaputra in Bangladesh before flowing into the Bay of Bengal. Early in its drift to the sea, the river wends near but never quite reaches the Kathmandu Valley, which thirty thousand years ago was a lake. Either the steady lapping of waves or a sudden tectonic shift sawed open its shoreline, draining the water and exposing a vast cradle of fertile land into which a metropolis of millions would take root.

Mel and I wandered in a daze through frenetic streets lined with temples and buildings and shops dealing in things we mostly did not need, but a few we did, such as visas for India and new bike tires. It was disorienting not to see mountains anymore. Although the Nepali capital is technically close enough to the Himalaya to reveal them, the range is redacted by smog in Kathmandu, as if to protect city dwellers from the fact of their smallness. "A mountain always practises in every place," observed the Buddhist poet Dōgen, but I have never been so enlightened as to not require the real deal. Sometimes I worry that I'm not wild myself, but terribly prone to tameness. I don't just appreciate huge, head-clearing horizons; I need them like a crutch, the sort of hard contours I can grab onto and heave myself up with to behold the vastness out of which we came and to which we will all return. Or at least get some fresh air, which in Kathmandu was in short supply.

What the city offered in abundance was food, but something I'd gorged on at the China-Nepal border had made me sick to my stomach, meaning the only calories I could handle were the same plain noodles I'd subsisted on across the Tibetan Plateau. This was cruel gastronomic punishment when the world's cuisines were suddenly on offer: Thai curry, Indian samosas, German cake, French croissants—you name it, Kathmandu had it. Whereas I had salmonella, or something like it, for the travel clinic's diagnosis proved inconclusive, and not just in terms of what ailed me. "Mr. Kate Harris," began the email with test results I'd subsequently receive weekly for the next six months despite my replies clarifying that (*a*) I was a Miss, thank you very much, and (*b*) I'd already received this notification. But the test results kept coming, bureaucracy itself the infection.

We set off a week later with the sun at our backs. After pedalling east across the Caucasus and Central Asia, and then south across Tibet, we would head west now across the Indo-Gangetic Plain of Nepal only to veer north again in India, back toward the Himalaya, spiralling in on the Siachen Glacier like a spaceship wooed into orbit by a planet's gravity. As we left the basin of the city, that same weak force worked against us. We struggled up Kathmandu's steep outskirts, and the valley seemed a lake still: our faces and clothes looked soaked, as though we were back on the Black Sea, only here it rained hot and from the inside out.

According to Buddhist lore, it wasn't tectonics or erosion that drained the Kathmandu Valley but the bodhisattva Manjushri, who saw a lotus growing in the centre of the lake and cut a gorge to let the water out, so that people could farm the fertile land below. Manjushri is typically associated with insight, but this wasn't such a bright move, for the former lake bed cradles seismically unstable clays and soils in one of the most tectonically volatile swaths of the planet. We left Kathmandu shortly before a small but still deadly earthquake rattled the place, though we didn't even learn about it until a week later, when we checked email and saw dozens of worried messages from friends and family. Four years later a 7.8-magnitude earthquake would shudder across the country, killing nearly ten thousand people and injuring tens of thousands more.

Even in the brief moment of relative stability that we biked in, the city seemed on the verge of falling apart. We pedalled past rickety apartment buildings out of which ganglia of rebar broke off like unfinished thoughts. Flimsy balconies looked structurally reliant on the lines of laundry strung across them, colourful T-shirts and saris flapping like urban prayer flags. The roads were just as vibrant and precarious, for traffic consisted mostly of Tata trucks painted with orange, red, and green patterns and exuberant slogans: "Road King," "Big Boss," and "One Mistake Game Over!" The trucks grunted out black clouds of exhaust as they heaved up the road, looking dangerous but also slightly comical, the vehicular equivalent of hippos. When their drivers saw us they honked their horns, a shrill, surprisingly girlish giggle of notes. I could tell how far ahead or behind me Mel was on her bike by the sound, the way you predict the distance of a storm from the seconds elapsed between lightning strikes and thunder.

At the turnoff to Pokhara, a popular trekking town, vehicles were backed up in both lanes for three miles. The atmosphere was remarkably calm for a traffic jam, even festive. Drivers napped on bamboo mats in the shade cast by their truck engines, and enterprising vendors wandered among the stalled vehicles selling fresh coconut slices and vanilla ice cream in neon orange cones. We weaved our bicycles through the jam, removing panniers when necessary to squeeze past vehicles parked too close, including a bus on which a kohl-eyed child stared at us out a window. When Mel smiled and waved, the girl looked terrified. "I'm really great with children," acknowledged my friend.

We didn't see any kids in Narayangarh a few days later, or anyone at all. The city was eerily deserted when we arrived, with steel garage doors pulled down on all the shops. The sidewalks were empty but for a lone woman whose bare, bony legs twitched sporadically, as if electrified. She stared nowhere in particular, a frantic look on her face. A metal begging bowl beside her collected nothing but sunlight. We continued toward the main intersection, where a massive crowd was chanting something we couldn't understand. Police in riot gear shuffled nervously around them.

"Politics," said the owner of the guest house we found, when I asked what was going on. Someone had been murdered, he explained, and his supporters were calling for the killer's execution. Maybe the pervasive unease in the city kept me awake that night, the sense that we stood on so many kinds of shaky ground, but the heat and humidity didn't help, and neither did the unsteady hum of a fan stirring the sludge of the air around the room. At some point I got up for water and glanced at the window just as a gecko walked up the far side, its tiny digits gripping the screen, its belly pearled in the beam of my headlamp. Outside, the mob was long gone but the woman was still there, a pile of sticks on the sidewalk. I could only make her out indirectly, by looking at the slightly lighter darkness surrounding her, the way you see stars best if you stare at the black sky between them. She didn't move, not even twitch, in all the time I studied the emptiness around her. I took a shower fully dressed and managed to fall asleep before the dampness fled my clothes.

Relief in Nepal is a rainy day. A day without shadows, a day of storms. It was drizzling when we arrived in Lumbini, a flat yawn of fields and forests unremarkable in every respect but one: Siddhartha Gautama, the future Buddha, first cried for his mother's milk here. According to legend, Siddhartha was born a prince in Lumbini more than two millennia ago. He was raised in opulence and shielded from all forms of sadness and despair. Only at the age of twenty-nine did he venture outside the palace gates, into the real world, where he was so shaken by the suffering he saw that he renounced his wealth and royalty and went wandering in search of a universal cure. After five years he found enlightenment somewhere between severe asceticism and sensual indulgence, by recognizing desire as the source of all suffering and devising a systematic way to appease it. Although the Buddha only became the Buddha because he fled this place and the life of privileged ease it represented, Lumbini is now a World Heritage site, its ruins and shrines and gardens all places of pilgrimage.

When we arrived, a group of Indian women in bright saris were huddling beneath umbrellas as they burned offerings of incense, the smoke rising between lines of rain. Several monks in saffron robes sat beneath a Bodhi tree, whose massive branches reached out for a dozen feet in all directions, then reached farther still thanks to the prayer flags unfurling from their leafy tips like feelers. One of the monks ushered me over and wrapped red string around my wrist, a blessing for the road. Nearby a few rickshaw drivers napped in the shelter of their back seats, waiting for passengers. When one of them importuned us to take a ride, Mel pointed out that we'd biked there. "So you must rest now!" the man exclaimed.

Water sprayed up between the wheels like jet streams. Being the baggage on a bike for once, rather than its engine, was rather enjoyable, plus the rickshaw came with a roof. We drove past a giant statue of the Buddha sitting in meditation, glowing a dim gold through the rain. I idly wondered how deep the precious element went on the statue. Was it solid gold, like the *paiza* Marco Polo reportedly carried, a tablet that served as a diplomatic passport within Kublai Khan's empire? Or just thinly glazed with the metal, like the Golden Record on the Voyager spacecraft, which is mostly copper? The cover of the record is aluminum, and it is etched with clues for making sense of its contents and origins: renderings of a phonograph record and the stylus needed to play it, a map showing the location of our solar system in relation to fourteen pulsars with distinct periods, and more. Electroplated over this is an invisible veneer of uranium, a radioactive element that would naturally decay into daughter isotopes over time, giving whatever scientifically literate extraterrestrial civilization that discovered the Voyager probe a sense of the eons elapsed since its launch.

Maybe the stories we tell ourselves, our cherished slant on things, is similarly vulnerable to slow decay. The farther I travelled along the Silk Road, the more the Golden Record seemed a lie, and not just because its surface didn't reflect its depths. Far more revealing than what was included in that cosmic-message-in-a-bottle, I'd come to realize, is what Sagan and his committee left out, namely details like women twitching alone on sidewalks. In fact, the record doesn't contain any hint of war or greed or death or cruelty. The closest it comes to acknowledging the existence of sadness is the sound of a baby crying, and its most squeamish inclusion is a picture of an ichneumon wasp, which was selected to illustrate the natural phenomenon of flight. An odd choice, given this parasitic insect burrows inside other insects to deposit its eggs, which hatch into larvae that gnaw their way out. "Other insects, like bees, have lives more in keeping with our moral and social sense," admitted Sagan, "but this creature is an inhabitant of Earth too, and who were we to pass judgement on its way of life?"

Yet Sagan's committee passed countless judgments on what was meaningful about life on Earth, and ultimately decided to present, with few exceptions, our best and brightest side to the cosmos. As a result, the Golden Record reads like a sanitized encyclopedia of earthly existence, the equivalent of the abridged version of Marco Polo's travels that I read as a kid. To their credit, Sagan's committee did include Blind Willie Johnson's haunting blues rendition of "Dark Was the Night," an old hymn about facing nightfall with no place to sleep but the cold, hard ground. Not that you can glean this from the lyrics, for there are none in this version: Johnson plays the bottleneck slide guitar and hums wordlessly through the three-minute track. All the wonder and heartbreak in the world brims over what he doesn't say, but I doubted alien beings would pick up on such subtleties. As such, the record effectively portrays the Earth as a planet whose peaceful denizens are devoted to music, flight, the art of possibility. The sort of cozy, blinkered world into which Siddhartha Gautama was born. Which makes it all the more remarkable that he left it behind.

The rain left snail-traces of slime everywhere it fell, including on the snails. When the tour was over, the rickshaw driver dropped us at a café to wait out the downpour. An ox had a similar idea and shuffled beneath the awning of the neighbouring fruit and vegetable market. Huge bones poked at the taut tarp of the creature's skin. A hump the size and shape of a human head stared out of its back, keeping watch while the ox nuzzled at limes stacked in neat green pyramids. The shopkeeper appeared with a pail and tried to shoo the creature away by flicking drops of water at its face—the real face, not the hump, but either way it was more a baptism than a deterrent. Finally the ox ambled off, a slow, unhurried swaying that set its hump nodding in all directions, from side to side and up and down in that manner so distinctive of the Indian subcontinent, as if saying yes and no to everything at once.

After Lumbini I saw Siddhartha everywhere I looked in Nepal, the way you learn a new word and suddenly notice it all the time. There was Siddhartha Bank, where the fat paunch of the teller sweated a dark halo through his shirt as he counted out rupees. There was Siddhartha Highway, paved in speed and ambition and occasionally flattened cobras, which looked from a distance like cracks in the road. Advertising in Nepal seemed universally energized by the belief that branding could redeem any enterprise, make it sacred by association or at least draw in devout customers. What would the real Buddha say, I wondered as I biked past Siddhartha Internet Café, if he could see such monuments in his name? Perhaps he'd remind us that the slower you travel on the

Siddhartha Highway, the more meaningful the trip. Maybe he'd stress that money is a false idol, a mirage induced by the heat differential between our heads and hearts. Or perhaps he'd giggle and say, *No problem*, as the Dalai Lama reportedly did when asked what he thought of the bars and nightclubs the Chinese had built near the Potala Palace. I suppose spiritual life, like any kind of travel, hinges on the everyday even as it taunts the trivial: the material world matters, the Dalai Lama implied by his answer, but only so much.

Farther west the roads were flat and full of bicycles, many as burdened or more burdened than ours. Some carried racks of live chickens swinging upside down by their feet, others hauled mountains of half-ripe bananas. The single-speeds popular in Nepal had high handlebars that forced their riders into extreme upright postures, lending a regal dignity to the pelotons. These bicycles were not just a means of transport but also of self-expression, whether in the form of bouquets of flowers fastened to handlebars or bright fabrics fluttering from hubs. Other cyclists stood out by their creative riding styles, like the little girl who was too small for her bike so she rode it *through the frame*, with one leg threaded between the top and bottom bars, standing in the pedals and hustling along as though bicycles were designed to be ridden that way.

White butterflies drafted off us as we biked toward Bardia National Park. At one point I pedalled through a cluster of them, and my wheels seemed to rise in the release of wings and dust from the road. We took a few days off at the park in hopes of seeing Nepali wildlife that wasn't roadkill. Despite the heat and humidity I wished I'd worn long pants and sleeves as we followed two guides into the Babai valley, an ecosystem evolved to impale: tall grasses so sharp it was like strolling through knives, and crocodiles lazing along rivers with sawblades for smiles. I kept mistaking water buffalo for rhinos, but we didn't see the latter until the next day, when a mother and her baby materialized on the far bank of a river. I don't know how long we watched them, but it was long enough for the fresh mud on their hides to go from shiny to a dull plaster. This species of rhino—the greater one-horned, or *Rhinoceros unicornis*—only survives in a few isolated nature reserves due to habitat loss, sport hunting (now banned), and poaching throughout its former range in the Indo-Gangetic Plain. Because of their use in traditional Asian medicines, rhino horns fetch high prices on the black market despite the fact that they are composed of pure keratin. Sickly people would benefit just as much, therapeutically speaking, from gnawing on their own fingernails. I could barely make out the mother rhino's miniscule horn, and the baby didn't have one, just a pair of eyes like wet pebbles, staring at us across galaxies. We spend millions of dollars trying to communicate with extraterrestrial civilizations, I marvelled to myself, but we still don't know what a rhino says when it snorts and sidles away in the grass. At least I didn't, though our local guide certainly did. "Time to go," he said, motioning for us to backtrack.

Reeds razored my bare legs as we hiked back to the lodge. On the way, the guide explained how monkeys sound alarms when they spot a Bengal tiger, which warns away the deer those tigers like to hunt. In return, however unconsciously, the deer nurture lice that monkeys harvest from their fur. We heard monkey alarms, but they were probably triggered by us, because we didn't see deer or tigers. What we did see was a hole the shape and size of a large fence post, dimensions I didn't even recognize as a possible footprint until our guide explained who made it. Elephants are among the smartest, most empathetic creatures on Earth, with brains as neurally complex as humans'; they use tools, bury their dead, shed tears, and occasionally go on furious rampages. The guide described how elephants sometimes ransack the rice fields near Bardia, ruining the harvest and crushing whatever mud-and-thatch houses stand in their way. And who could blame them, with humans encroaching on their territory on all sides? An electric fence had been set up around the perimeter of the park to protect local villagers, but the fence only worked when there was power, and the power went out often, as we learned that night after dinner.

When the lights went out Mel retired to our room, but I lingered in the dining hall, reading by headlamp and a candle that sizzled brighter whenever bugs flew into the flame. In the lodge's small library, among the usual mysteries and thrillers and romance novels that pile up in places frequented by budget travellers, I'd been surprised to find a book about Milarepa, the Tibetan Buddhist monk and poet who lived in a cave and subsisted on a diet of nettles, which seemed less a path to awakening than to madness and starvation. I wanted badly to understand Buddhism but couldn't seem to wrap my mind around its most basic precepts. Didn't a desire for enlightenment still count as desire, meaning the goal is undone by the longing to achieve it? And what was so compelling about nirvana anyway? Wasn't permanent departure from the cycles of rebirth just a glorified way of checking out, effectively a one-way mission to Mars? The wish to flee *samsara*, the phenomenal world with all its flaws and illusions, smacked to me of escapism. Give me this beautiful, broken planet over any kind of blank slate. Then again, I suppose our own flaws and illusions are the hardest to see, never mind flee.

Outside, cicadas clattered metallically in the dark, a chorus of spilled cutlery. I made my way to the room by headlamp and thought about Siddhartha's mad wish to end the suffering of all beings: you couldn't accuse the guy of aiming low, settling for the possible. And why not err on the side of audacity when it comes to this one and only life? Of course, Buddhists beg to differ about the one and only part, but still, taking no chances, exploration seemed the most rational possible response to finding yourself alive. You wake up suddenly in a strange land: What else is there to do but look around? Explorers might be extinct, in the historic sense of the vocation, but exploring still exists, will always exist: In the basic longing to learn what in the universe we are doing here. In the sort of "absolutely unmixed attention" the philosopher Simone Weil called prayer. In any small, useless, searching gesture—a pilgrimage, a poem—that admits suffering even as it tips the balance of the world, however briefly and immeasurably, toward wildness.

Mel was fast asleep and the room was dark. I brushed my teeth by headlamp and lay down on a mattress that still held the heat of day. When I dozed off I dreamed of rhinos with distant nebulas in their stare, and tigers padding through the forest with full-moon eyes—lives flawed and restless and full of desire, which is another way of saying full of stars.

The true risks of travel are disappointment and transformation: the fear you'll be the same person when you go home, and the fear you won't. Then there's the fear, particularly acute on roads in India, that you won't make it home at all.

At first traffic was quiet just beyond the India-Nepal border, the final frontier of our trip. Then we saw cars swerving erratically on the horizon, as though the drivers had lost their minds or were dodging enormous potholes. In fact, they were dodging monkeys. A hooting, scratching, shuffling congress of primates occupied the pavement, several with babies gripping the fur of their arched backs. Unnervingly humanoid heads swivelled to watch us pass. The sudden, jerky movements with which we manoeuvred around these road pylons made us look deranged ourselves—a word from the French meaning "to flee from orderly rows," and the organizing principle for traffic in India.

The road became an obliterating press of dust, horns, people, and fumes. Cars teemed over the tarmac like flies on rot. The sky was sour and curdled-looking with pollution and heat, and the air smelled by turns or all at once like shit, curry, burning tires, woodsmoke, the sulphurous burst of matches flaring, urine brought to a boil, and some kind of chemical that singed my nose like swimming pool chlorine. Occasionally the blasting of a bus horn swept the congested road clear, a sonic plough, and the ringing in our ears was almost a relief because we couldn't hear the shouts of "Hello!" "How are you?" and "What is your name?"

Mel and I ignored these catcalls, but our temporary new travelling companion, Hana, took them for earnest greetings and never failed to wave back when someone blared their horn. An environmental lawyer friend from British Columbia who in her spare time builds guitars, Hana flew to India to join us for a few weeks after passing her bar exam and before starting at a law firm. Her idea of a holiday was pedalling all day, every day, through deadly traffic, constant ogling, choking pollution, and withering heat. It was admirable, really, Hana's openness to the experience, but I worried that her enthusiasm would prove wearying. In fact the opposite was true, for if anything, her frequent exclamations of "Holy cow!" alerted us to the shifting baseline of what Mel and I deemed bizarre. How had sacred heifers nuzzling at trash in a snarl of traffic become ordinary to us?

The three of us stopped for a break when we saw a sign for a Sweet Shop. Inside I picked out a candied green square that looked delicious, but when the shopkeeper handed it to me, I noticed a fleck of fly wing stuck to it. I pointed this out to the shopkeeper, who looked appalled and swapped it out for a different square—that had another fly wing on it. I pocketed the sweet when the teller wasn't looking and turned to address the mob of friendly young Indians who had surrounded us, making rapid-fire small talk. "How are you?" "Where are you from?" "What is your name?" On and on, in such quick succession that even Hana grew a little flustered. Eventually a hefty young man with an air of authority pushed himself forward and instructed everyone to quiet down. He introduced himself as Alok, and explained, in an urgent tone, that people were waiting for us next door.

"Sorry?" Mel said.

"Please, ladies, only five minutes. You must come," he implored. "They are waiting!"

Who was waiting? Why, and for whom? These are the great mysteries. Glad for any excuse not to get back on the bikes yet, or ever, we followed Alok to a nearby building that a sign identified as an English Institute. Inside, the walls were covered in posters with inspirational sayings like "Impossible Says I, Am Possible" and "Great Communication Is Demand of Universe." People were indeed waiting for us, with chai and samosas. These Indian students deployed their English questions at a more conversational pace in the moderating presence of their teachers, and we took turns answering them. "We are very well, thank you." "We are from Canada." "I am Kate, this is Mel, and this is Hana."

"How do you like India?" one girl asked earnestly.

Dozens of earnest faces crowded close, anxious to hear our answer. Mel looked to me to field the question, I looked to Hana, and Hana looked back to Mel. In the silence of us not answering there was a sense of gathering speed. Then Mel cleared her throat and turned to the girl.

"It is very hot," she offered diplomatically. The students laughed in agreement and relief.

Only in the Himalayan foothills did the world cool down again. Even the light was crisper, edgier, honed to a higher-altitude glint. The air smelled of pine and shade and mist, and I felt like I could breathe for the first time in a month. We cheated to get here— namely, we cheated certain death by skipping just over a hundred miles of heinous traffic by taking a bus to Shimla. This former British hill town is like a stone dropped in water: hazy waves of hills roll out in concentric circles from its perch of cobbled plazas and colonial buildings. Though Shimla isn't high by Himalayan standards, it is just high enough to take the edge off the heat. It is also where you have to go to travel any higher, for a special permit is required to visit the Jammu and Kashmir province of India, in which Ladakh is located.

Hana fixed her hair before posing for her Inner Line Permit photo, though Mel and I didn't realize it until we compared portraits afterwards. Despite posing in front of the same glaring white screen, Hana somehow emerged with a glamour shot, with soft light haloing her immaculately brushed locks, whereas "mug shot" was too delicate a term to describe Mel's and my photos. Then again, Hana had an unfair advantage: her "expedition" packing included four (four!) fancy-smelling soaps, deodorant, and outfits colour-coordinated to match the red paint on her bike and panniers. And whereas Mel and I only had one pair of bike shorts each, Hana brought two, though she didn't let on about this until a week later, when a monkey stole mine. At least I assume it was a monkey, because who else would filch a stinky pair of bike shorts hung out to dry overnight? Hana generously offered me her second pair.

Thank goodness we had a few things to tease Hana about, because she was bionic, turbo-powered, easily outpacing us up mountain passes as if she'd been the one biking the Silk Road for nine months. As I huffed along far behind her I reminded myself that she was carrying less—no cameras or laptop, a lightweight bivy sack instead of a tent—but I knew these were the barest of handicaps. Only when the pavement disappeared, and landslides began posturing as roads, did I have a slight technical advantage. On one of these rough stretches a pair of herders came at us shouting, "*Atcha!*" and "*Tigge!*" as they urged hundreds of horned creatures along what was little more than a gash in a mountainside. With a sheer drop to our left and a sheer cliff to our right, we had nowhere to go, but the herd parted neatly around us, human boulders in a river of musk.

We reached Kalpa just as the sun set, the glow slowly moving from the mountains to the stars. Although we weren't technically in Ladakh yet, and Hindu shrines were more common than prayer flags along the road, a Buddhist monastery in town was blasting a musical rendition of "*om mani padme hum*" from massive speakers. Religion is nothing if not loud in India, at least on the outside of temples. Inside a group of monks sat in contemplative silence. I badly wanted to take their picture but didn't dare disturb them. I shouldn't have worried: at one point a cellphone rang from somewhere deep in the folds of a robe on a wizened old monk. He fished it out and chatted loudly as his companions continued to meditate.

Beyond Kalpa the terrain grew so steep I flinched when a flock of birds flew above me, thinking they were falling rocks. Farther down the road a waterfall blessed every car (and cyclist) that went by. After four and a half hours of biking, which somehow took all day, we saw a sign for the town of Pooh. A look of joy bloomed on Hana's face. She dug into her bags, and a few seconds later held up Winnie-the-Pooh stickers.

"You just happened to have these?" I exclaimed.

"So prepared!" teased Mel.

A friend had given Hana the stickers, figuring they'd make a lightweight, portable gift to pass out to children along the way. For now we stuck them on the sign as pictographs. When Hana's holiday ended a few days later, we were sad to see her go.

The cold air sharpened itself on the mountains as Mel and I climbed farther north. Vast slopes of crumbling rock leaned into the light at the angle of repose. Switchbacks veered up and over contours so corrugated they made the road, from a distance, appear to break off suddenly at random angles. Horizons were as suggestive in some directions as they were solid in others, which was part of the appeal of the Himalaya—the way the landscape always keeps you guessing.

The same could be said for the Silk Road. Until now the end of this ride had seemed so far away it didn't bear thinking about, or at least I hadn't thought much about it, though Mel had been applying for jobs since Kashgar. She knew what she wanted to do and where she wanted to live: work for a community non-profit in Toronto. All I knew was that there was no going back to the lab, that I would never stop exploring. The French sailor Bernard Moitessier had the right idea: after taking a generous lead in the inaugural Golden Globe of 1968—a year-long, solo, non-stop, round-the-world yacht race—he was so content at sea that he skipped the finish line, forfeited the cash prize and world record, and kept on sailing. "Maybe I will be able to go beyond my dream," he mused, "to get inside it, where the true thing is . . ."

So what was inside the Silk Road, beyond it, where the true thing was? It wasn't that I wanted to bike on and on around the globe like an astronaut stuck in low orbit, falling and missing the earth forever. What I pictured at the end of the road was a log cabin insulated with books near the Juneau Icefield—an existence rich in mountains, words, stars, wildness, really everything but money, but when it comes to that, who needs more than enough? Lacking Darwin's inherited affluence, I figured I could eke out a basic living, like Wallace, as a freelance explorer, doing just enough paid work to buy time and space for my own reading, wandering, and writing. It wasn't a different world I craved anymore so much as this world done differently, done better for everyone. Maybe the right words could launch us there, or more accurately bring us home.

The road climbed above a river the colour of tiled domes in Samarkand, meaning ten thousand shades of turquoise. Feeding into it were streams too shallow to carry any colour, their water approaching the clarity of ice. One of them drained from a teardrop-shaped lake where, according to our map, a series of *dhabas*, or roadside restaurants, operated in the summer. The place was deserted now. A few bright paddleboats were still parked on the shore. "That's our sport!" exclaimed Mel. "We'll make *waves* with these legs." But the boats were padlocked and half-sunk with water. We set up the tent and cooked dinner instead.

The last light of day cooled on the mountaintops. We sat outside drinking tea in down jackets as it grew dark. I watched a nearly full moon float over the peaks, growing smaller the higher it got, and thought about Armstrong's first stroll on it, or more accurately what happened after. When he and Buzz Aldrin returned to the *Eagle*, the insect-like lunar module, they discovered a broken switch on the circuit breaker required to ignite the ascent engine. The astronauts reported the problem to Mission Control and tried to nap as the earth-bound engineers brainstormed solutions. Hours passed as fellow astronaut Michael Collins orbited above them in the *Columbia* command module, seventy miles away and completely out of reach. Whether it was inspiration or desperation that finally prompted Aldrin to jam a felt-tipped pen into the circuit breaker is hard to say. Either way, the makeshift switch worked, and so it was rockets but more crucially a pen that launched the astronauts off the moon and on a trajectory back to Earth.

A few days later we topped out on Baralacha La, a pass marking the edge of Ladakh. The road twisted down through flame-licked mountains, then a canyon carved by an emerald river. Eventually it flattened out across a mesa where yellow grasses flared in the setting sun. Herds of yaks seemed to graze on pure light. "No wonder you love Ladakh!" Mel said as we biked side by side. "It looks like Mars, only more alive."

Raised by the same crash and warp of continents that threw the Tibetan Plateau skyward, Ladakh in many ways felt more like Tibet than Tibet. I thought back to my first morning in Leh, the regional capital, when I travelled there on a summer break from MIT: the Dalai Lama had driven past my hotel, grinning widely out the window behind his trademark glasses. People lined the street into town and watched him pass in rapt silence, bouquets of incense smoking from their fists. Across the border, in Tibet, barely a hundred miles away, possessing a photo of His Holiness could get you arrested.

But more than the bleached prayer flags on every pass, the Buddhist monasteries barnacled on cliffs, the infrastructure of ice and rock and sky, it was the slant of light that brought me back to Tibet. It fell in huge, silken throws across the mountains of Ladakh, articulating the creases and folds of the land until it resembled the face of someone just woken, cheeks impressed with the pattern of some huge pillow's fabric. I couldn't stop rubbing my eyes.

Nothing like a motorsport rally to really jolt you awake. Mel and I were spinning up the first of the twenty-one switchbacks of the Gata Loops, a stretch of road that coils up a mountainside like a small intestine. I didn't think anything of the dirt bike that revved past, spitting up gravel. A few minutes later another nearly hit us on a hairpin turn. Then a souped-up jeep careened by, the smooth white helmets of its driver and passenger shaking like eggs in an out-of-control crate. It was followed by more than sixty dirt bikes, quads, and rally cars, for we'd inadvertently become the slowest contestants in Raid the Himalaya, the world's highest and toughest motorsport rally.

Hyped as "the ultimate test of man and machine," racers cover up to two hundred miles a day during the week-long event, which starts in Shimla and ends in Leh, though the route can vary from year to year. Drivers contend with non-race vehicles during early race stages, but the Leh-Manali Highway is closed to traffic as the race moves farther north. Mel and I weren't aware of the closure because we'd been camping between towns, and drivers weren't aware of us, for they assumed nobody else was on the road. This made for some terrifying near-misses, particularly on switchbacks, where it wasn't easy to scoot out of the way. "Ahh, serenity," sighed Mel as we pressed ourselves into a cliff to avoid death by yet another rally car.

The stop-start nature of our day meant we didn't cover the distance planned. Instead of biking four high passes, we gave up after the third, in a valley called Whisky Nala, where we hoped to find its namesake or at least some water. But the only stream was frozen, and the valley was abandoned except for a Raid the Himalaya checkpoint. After cursing the motorsport rally all day, I sheepishly asked the checkpoint volunteers for water. They handed me bottles just as the last competitor revved through, and then they drove away themselves.

Silence rang off the peaks. The sun sank behind the mountains and the air instantly felt twenty degrees colder. Mel and I put on all the clothing we had and crawled into the tent, then into our sleeping bags. It wasn't bedtime, but it was too cold to talk, to eat, to read. As I stared at the ceiling I thought about a Buddhist saying I'd read somewhere—"nowhere to go, nothing to do"—and marveled at how accurately it described our life on the road at times like this. Nowhere to go, and nothing to do but stay warm, stay still, wait for darkness to arrive like a bleed of ink from one side of a page to another.

We slept until the sun lit the tent, a slow fuse of heat. Everything was frozen: the water tucked into the bottom of our sleeping bags, our toothpaste, our noses and toes, even the bike lube, as Mel discovered when she tried to apply some to her squeaky chain. The stove was reluctant to light with so little oxygen, and when it finally sizzled to life, the ice took forever to thaw. We forced tepid oatmeal down our throats and set off shivering up a 16,400-foot pass.

Cracks in the road seemed to testify to something enormous surging up from below, a monstrous root system or new mountain range. Mel had strapped the empty Raid the Himalaya water bottles to the back of her bike, where they rattled and flailed and at one point flew off. "Nooooo!" she wailed. "My life savings!" I stopped to help her pick them up but wished I hadn't: ten seconds into pedalling again I felt the usual flush of deep fatigue in my legs, a crawl of acid through my thighs and calves. Mel felt the burn at the exact same moment, judging by our synchronized groans. The pain of starting again cancelled out any relief in stopping. We learned to inch along no matter the pace, even when the only force on the pedals was the weight of our legs pulled down by gravity and wearily hauled back up again. I couldn't remember how it was to feel fresh, legs firing strong, hands gripping the handlebars firmly rather than hovering weakly above them, my fingers barely making contact because my wrists were so sore from the jackhammer roads.

Everything was falling apart, grinding down. Our clothing was in tatters, our socks poked through holes in our shoes, my watch strap was broken and the battery dead. Mel's bike had been missing a handlebar grip since Kyrgyzstan, a bungee cord was tangled permanently around her rear hub, and my kickstand and mirror had bailed on me several countries ago. Our inner tubes were more patches than rubber by now, and the chains and gears on our bikes protested loudly with every pedal stroke. The tent zippers refused to zip, the bit of wire that fixed the burner to our stove was broken, and our Therm-a-Rest mattresses leaked from pricks so microscopic we couldn't find and plug them, so we woke up hugging the cold, hard ground. After nearly a year on the road, it was a wonder anything still worked, especially our friendship.

That was saved by the smallest of things: The way Mel would kick out her legs with a goofy flourish when I glanced back to see where she was. Her sense of absurdity perfectly matched the landscape we passed through, and her commentary about it was always illuminating, such as when we came across a patch of ground creepily strewn with long grey hairs. "Satanic rituals involving shears and senior citizens," confirmed Mel. "You said that a little too knowingly," I replied. Both of us craved solitude as much as company, and more than anything it was this ability to be alone together that let us survive the trip. Of course we had bad days, terrible days, like when Mel poured herself a hearty mug of instant coffee at breakfast and drank it all before noticing she'd barely left me any hot water. "Oops!" she said, as if morning coffee weren't a matter of life and death. "You grew up with siblings, Mel!" I fumed. "You know proportions *matter*!" I drove her crazy at times too, but even so she'd follow me to the ends of the Earth and I'd follow her in turn, because that's what we did on a daily basis. We biked on and on, whatever our mood or the weather, until finally, after ten months, we were as close as the Silk Road gets to the stars.

Maybe I was just woozy from altitude, for at 17,480 feet Taglang La was the high point of our trip, but I swear I saw the curvature of the planet from that pass. I felt convinced no bicycles had ever flown so high or so far. Of course this wasn't true: others have biked higher and farther, and certainly faster, with fewer flat tires and false turns. But exploration, more than anything, is like falling in love: the experience feels singular, unprecedented, and revolutionary, despite the fact that others have been there before. No one can fall in love for you, just as no one can bike the Silk Road or walk on the moon for you. The most powerful experiences aren't amenable to maps. Nor are they amenable to words, at least when you're too out of breath to say much at all. Mel and I tied a string of prayer flags to the tangle of others on the pass and let them do all the talking.

We coasted back to Earth for hours upon hours through a landscape almost lunar in its starkness, a magnificent desolation of broken purple and red peaks. Halfway down the pass a white-washed monastery gleamed high on sunlit cliffs, and the walls of the village below it were covered in neat wads of dung left out to dry in order to fuel winter fires. Poplars dropped their alms of leaves along the road, more gold than coins can ever be. We turned into a canyon formed by huge plates of rock that sliced up through the crust like dinosaur vertebrae. The river the road followed ran apricot or copper or gold, depending on the mood of the rock it reflected. Above a river of sky mirrored the water's course through steep mountains.

When we finally spat into the Indus river valley, our lungs full of oxygen, I was half-tempted to bike the last thirty miles to Leh. But why rush to the end of the road? Instead we stopped at a restaurant in Upshi for momos, a kind of dumpling, then we pitched the Glow-worm for the last time on the bank of the Indus, that rush of meltwater straight off Siachen.

In some ways it was the closest we'd get to the glacier. We biked into Leh the next morning, and just in time: within days a blizzard blocked the high passes we'd just pedalled across, closing the road for the year. A different road to the Nubra Valley and Siachen remained open, but we needed permits to go there, and those would take a few days to process. To pass the time, we took a shared jeep to the Indian shore of Pangong Lake, that spill of turquoise water we swam in in Tibet five years earlier.

The edge of winter, the edge of the Tibetan Plateau. Mel and I stood shivering in the spot we would've landed if we'd kept swimming east that first summer on the Silk Road, a faint slick of sunscreen in our wake. Then again, shortcuts never take you to the same place. Wearing down jackets and pants with the legs rolled up, we shuffled into water so calm and clear it was like wading through air. Ten seconds later we shuffled out again, numb from the shins down. That night we warmed up in the village of Spangmik over a dinner of dal-and-rice with two Indian tourists. All I remember from our conversation was that the men hailed from some massive city, Mumbai or Calcutta, and Pangong Lake was the first place they'd seen stars.

Back in Leh we picked up our Nubra Valley permits, then continued by car to Panamik, the last civilian outpost before Siachen. The driver let us out at the final checkpoint we were allowed to see, more than fifty miles from the ice. We walked toward the lowered guardrail as a cold wind rattled the chains that secured it, then stopped and stared at the end of the road. I don't know why I thought it would be any different now, the glacier miraculously more accessible, when nothing had changed politically. Soldiers from both sides still lived year-round at absurd heights, fighting avalanches and altitude sickness, wearing the same white camouflage and speaking essentially the same language, like a unified army. Meanwhile all around them, shimmering like mirages, were mountains trespassed by borders that nations swear have been there all along.

The sun torched the rims of the peaks in Panamik. A flock of birds folded and unfolded the sky. More cold gusts stripped the poplars of the few leaves they had left, the wind more alive than the branches it moved, and so big it could only be the mountains breathing. Mel headed back to the car, hands in her pockets. I took one last look in the direction of the glacier, not so much to catch a glimpse of its ice but to give its wildness my full attention, if only for a moment.

Then I turned around. In the end Siachen, like Mars, wasn't a place to reach but a reason to go.

EPILOGUE

*D*eparture is simple: you step out the door, onto your bike, into the wind of your life. What's hard is not looking back, not measuring gain or loss by lapsed time, or aching legs, or the leering mile markers of ambition. You are on your way when you decipher the pounding of rain as Morse code for making progress. You are getting closer when you recognize doubt as the heaviest burden on your bike and toss it aside, for when it comes to exploring, any direction will do. You have finally arrived when you realize that persistent creak you've been hearing all this time is not your wheels, not your mind, but the sound of the planet turning.

I watched people spill off a crowded bus in Leh and line up in a neat row, presumably to use a restroom. I was wrong. They were waiting to spin the giant prayer wheel near the city gate Mel and I had biked through a few days and forever ago. Several women wore their hair in long braids that looped together behind their necks like reins. Bundles of hay winged the backs of some men. Each sun-hewn face was creased with smile lines—the map of a hard life with certain redeeming hilarities, such as the dog sprawled out directly under the prayer wheel, soaking up all the good karma being released. I'd seen other dogs under similar prayer wheels across Ladakh, for they recognized a safe haven, or at least shade, when they saw it. The men and women were careful not to step on this dog's tail as they spun the wheel clockwise, adding their momentum to its mantras.

Mel and I were wandering separately around the city, craving solitude and buying souvenirs, our exhaustion beautifully earned. She was making travel arrangements for her boyfriend to join her on holiday after I flew home, and we'd made plans to meet at a tea house later that afternoon. I arrived first, the only customer there, so I got out the laptop and started editing photos to pass the time. Faint lights cast shadows on the cold concrete floor, yet there was a coziness to the place, with candles on every red plastic table and a large poster of the Potala Palace on the wall. Eventually a stout, motherly woman with a face like softened butter approached to take my order. I quickly glanced over the menu: among the items on offer were momos, tsampa, and "yuck" butter tea.

"I'll have a pot of honey-lemon-ginger tea, please."

"Oh nooooooooo!" the Ladakhi woman howled. She dashed to the front of the tea house and yanked down the corrugated metal awning, then switched off all the lights. I sat in the dark, mildly alarmed, wondering what I'd said or done to provoke this.

"We forgot! Hee heeee!" the woman giggled, her voice disembodied in the dark. I heard the scrape of a match, then light flared above a candle. She waved for me to join her at the front of the shop.

"Look, look," she urged, pointing at an empty bolt hole in the awning.

I peeked through the pinky-sized gap: it was like squinting into the viewfinder of a microscope or a reproduction of Galileo's telescope, and I half-expected to see the rings of Saturn, *Rhodospirillum rubrum*, or the sign for the King's Arms pub. Instead I saw the street flowing with people, some holding signs, others candles, a silent river of flame.

"It's for Tibet," she whispered.

The woman and her husband had forgotten about the march for Tibetan solidarity, she explained, so they were inadvertently playing hooky. For reasons I couldn't understand, it was better for them to pretend they were away than to show up late.

"But can I go?" I asked.

"No, Miss, they will see!" the woman insisted. "Sit, sit, drink tea."

She brought me a steaming metal pot and some honey. I showed her a photo from Tibet on the computer, explaining that my friend and I had just biked there. The woman sat down, riveted. "More?" she asked.

I hesitated to share the androgynous face masks and Chinese flags that Mel and I had used as disguises. "This is good, so smart," she commented, giggling in conspiracy. Relieved, I moved on to red Chinese flags on traditional Tibetan homes. "They were made to," she said with a sigh. "They had no choice." She brightened at photos of power lines snaking along the highway—"Good, this is good"—and I shouldn't have been surprised but I was. When I showed her the pilgrims prostrating themselves to Lhasa, the woman murmured something I couldn't make out. At photos of Chinese tourists on high passes, where plastic bags fluttered among prayer flags, she clucked her tongue and was silent for a moment. Then she said, very softly, "Chinese government very bad. But

Chinese people not bad. They have same problems as Tibetans."

With that the woman disappeared into a back room, leaving me stunned at her refusal to take sides. She returned a minute later with some photographs of her own: A family snapshot featuring rows of solemn people wearing dark robes with sleeves so long they hid everyone's hands. A monastery pearled among gritty mountains. Some kind of Buddhist painting, intricate curves and symbols and patterns rendered in yellow, green, red, white, and blue.

"Sand," the woman clarified. "This is sand."

I'd read about how Buddhist monks painstakingly arrange bits of coloured quartz into a geometric representation of the universe, or mandala, then scatter the art in a gesture of non-attachment. The photograph I held was the sole proof that the sand mandala had ever existed, only the real mandala wasn't the completed work of art, but its attempt. That act of pure attention, the motion there and away.

The husband chuckled next to me. He was clicking through the photos on my laptop now, and he'd found one of Mel high-fiving a statue of a Chinese police officer.

After finishing my tea I packed up my things, and the woman lifted the metal awning to let me out. By now the march was over, the streets dark and empty, except for a wobbling light in the distance that I guessed might be Mel, heading back to the guest house after finding the tea shop closed. I shouted her name but the figure disappeared around a corner. I clicked on my headlamp and walked in the same direction.

The air was so cold my teeth ached. Snowflakes accelerated into the light and disappeared in the enveloping dark. A pack of stray dogs howled across the city, hymns freighted with burrs and distances. In a few hours, before dawn, the muezzin's call to prayer would sing across Leh, shrill and heraldic, followed by the low thrum of Buddhist long horns. In a few days a flight attendant would ask if I preferred steak and rice or beef and noodles for my meal, and I'd laugh and barely resist shouting *"beefandnoodles, beefandnoodles"* as nostalgia overwhelmed me—though not exactly nostalgia, with legs still too sore for that, but I wouldn't know what to call that species of longing for a random night in Uzbekistan. For the hungry days when Mel and I lived on instant coffee and laughter and scraps of light, and lived well. And in a few months I'd move off-grid with someone I love to a cabin in Atlin, near the Juneau Icefield, where I'd travel the Silk Road in sentences over and over again, and only gradually come to understand where I'd gone.

Of course I knew none of this as I wandered through Leh, as lost as I'd ever been in my life. My headlamp was almost dead, so I turned it off and found my way forward by looking up, walking in the faint gap between where the walls of buildings ended and deep space began. The dogs quieted and for a moment I heard whale song, a baby crying, Blind Willie Johnson humming the blues. Then silence, the hush of snow rewriting all the roads.

ACKNOWLEDGMENTS

"There is no such thing as a solitary polar explorer," observed Annie Dillard, and the same is true for writers. Every page of this book and the experiences it describes were made possible by the kindness of the strangers who befriended and helped me in my travels; the teachers who encouraged my addiction to questions; and the authors who inspired me out the door and eventually back to the desk. I have so many people to thank.

Foremost among them is Mel Yule, dear comrade in exploration, who since the age of ten has propelled me to places I never would've dared alone. Creighton Irons, Laura Boggess, and Jesse Stone Reeck infused my time at Carolina with goofiness and soul. The Jar Kids, Marcie Reinhart, Mike Moleschi, and Jamie Furniss made Oxford a place of magic and mystery. Sara Bresnick, Linnea Koons, Andrew Frasca, and Alex Petroff kept me company on long bike rides or in the lab at MIT. Lori Ormrod, Bernadette McDonald, and David Roberts somehow believed in this book before I'd written a sentence. Sarah Stewart Johnson has bolstered my conviction that we're not alone in the universe. Alison Criscitiello and Rebecca Haspell will always be my Fanny Pack. Thanks to my friends in the north—especially Wayne and Cindy Merry, Philippe and Leandra Brient, Dick Fast and Maggie Darcy, Judy Currelly and Stephan

Torre, Don Weir, Oliver Barker and Piia Kortsalo, and Cathie Archbould and Jacqueline Bedard—for supporting the writing of this book, not least by wooing me away from it with hikes and home-cooked meals. Thanks also to Libby Barlow for the cabin at the end of the road.

The Morehead-Cain and Rhodes scholarships both widened my world in ways I can't possibly express, but I hope this book is a start, as well as a token of my gratitude. Heartfelt thanks to Seven Cycles, Polartec, WINGS WorldQuest, OneWorld Sustainable Investments, The Wild Foundation, and everyone who supported the Cycling Silk expedition, with shout-outs to Milbry Polk, Vance Martin, Ruthann Brown, and Berna and Diarmuid O'Donovan. Working for the Earth Negotiations Bulletin team has been a regular source of inspiration and solvency; special thanks to Kimo Goree—I still owe you a bike ride.

The Ellen Meloy Desert Writers Award, Banff Mountain and Wilderness Writing Program, British Columbia Arts Council, and Canada Council for the Arts afforded me the means and time to write. Marni Jackson, Tony Whittome, Fred Stenson, Lori, Kim Rutherford, JanaLee Cherneski, Erin Fornoff, Elizabeth Reed, Karen McDiarmid, Tanya Rosen, and Mel read early chapters or drafts of the manuscript and made it vastly better. Any flaws or inaccuracies that remain are, of course, all mine. Thanks also to Doug Carlson and Stephen Corey for publishing an essay in *The Georgia Review* that eventually expanded into this book. Writing retreats were generously offered to me by Karen at Shawnigan Lake, Mel Ashton and Chris Pleydell at Ségur-le-Château, Cathie and Jacqueline at Lina Creek, and Jan and Pat Neville in North Carolina.

Thanks to my wonderful agent, Stuart Krichevsky, for helping me tease one possible Silk Road out of many and travelling it with me to the end. Deep gratitude to my publishers, Anne Collins at Knopf Canada and Lynn Grady at Dey Street Books, for their faith and enthusiasm. I'm grateful to Ross Harris for launching this book into lands abroad; Rick Meier for handling permissions and proofs with skill and aplomb; Five Seventeen for his dazzling design work; and Deirdre Molina, Ruta Liormonas, and Libby Collins for all they do behind the scenes. I'm especially grateful to Amanda Lewis, who made the writing itself an experiment and adventure, and to Lynn Henry and Matthew Daddona, who have been such fierce champions for this book from the beginning.

And finally, thanks to my parents, brothers, and extended family, who truly believed (feared) I'd make it to Mars. I'd choose a sheep shed with all of you over a new world any day. For Kate Neville, my love is as deep as Sloko Inlet on a late summer day, the wind quiet and the lake calm, paddling home on meltwater, mountain light.

PERMISSIONS

Excerpt from *The Writing Life* by Annie Dillard © 1989 by Annie Dillard. Reprinted by permission of HarperCollins Publishers.

Excerpt from *A Field Guide to Getting Lost* by Rebecca Solnit © 2005 by Rebecca Solnit. Used by permission of Penguin Random House Limited.

Quotation from *In the Skin of a Lion* by Michael Ondaatje © 1987 by Michael Ondaatje. Reprinted by permission of McClelland & Stewart, a division of Penguin Random House Canada Limited.

Quotation from *Notes from a Bottle Found on the Beach at Carmel* by Evan S. Connell © 1962, renewed 1990 by Evan S. Connell from *Notes from a Bottle Found on the Beach at Carmel: A Poem.* Reprinted by permission of Counterpoint.

Quotation from "The Clearing" by Tomas Tranströmer, translation by Robert Bly © 1962 by Robert Bly from *The Half-Finished Heaven: The Best Poems of Tomas Tranströmer.* Reprinted by permission of Greywolf Press.

Excerpt from *The Anthropology of Turquoise* by Ellen Meloy © 2002 by Ellen Meloy. Used by permission of Penguin Random House Limited.

Quotation from "The Swimmer" by John Cheever © 1947 by John Cheever, renewed 2000 by Mary W. Cheever from *The Stories of John Cheever.* Reprinted by permission of Penguin Random House Limited.

SELECTED BIBLIOGRAPHY

Works quoted, consulted, or referred to in the writing of this book.

EPIGRAPHS

"To speak of knowledge . . ." in Virginia Woolf, *The Waves*. London: Vintage, 2004.

"How we spend our days . . ." in Annie Dillard, *The Writing Life*. New York: Harper Collins, 2009.

"Never to get lost . . ." in Rebecca Solnit, *A Field Guide to Getting Lost*. New York: Penguin, 2006.

"I should like to do . . ." Ellen Meloy, *The Anthropology of Turquoise*. New York: Vintage, 2003.

PART ONE

1. *MARCO MADE ME DO IT*

Thesiger, Wilfred. *Arabian Sands*. Harmondsworth: Penguin, 1974.

Cherry-Garrad, Apsley. *The Worst Journey in the World*. New York: Carroll & Graf, 1997.

Nansen, Fridjof. *The First Crossing of Greenland*. Cambridge: Cambridge University Press, 2011.

Service, Robert, *Songs of the Sourdough*. Toronto: William Briggs, 1908.

David-Néel, Alexandra. *My Journey to Lhasa*. Boston: Beacon Press, 1993.

Rugoff, Milton. *Marco Polo's Adventures in China*. New York: American Heritage Publishing Co., 1964.

Yule, Henry and Cordier, Henri. *The Travels of Marco Polo: The Complete Yule-Cordier Edition, Vol I and II*. London: Dover Publications, 1993.

Moule, A.C. and Pelliot, Paul. *Marco Polo, The Description of the World*. New York: Ishi Press, 2010.

Polo, Marco. *The Travels*. London: Penguin Books: 1974.

2. ROOF OF THE WORLD

"Longing on a large scale . . ." in Delillo, Don. *Underworld*. New York: Simon & Schuster, 1997.

Harrer, Heinrich. *Seven Years in Tibet*. New York: Penguin Putnam, 1996.

Hilton, James. *Lost Horizon*. New York: Simon & Schuster, 1939.

3. NATURAL HISTORY

Shakespeare, William. *The Tempest*. London: Penguin Classics, 2015.

Kuhn, Thomas. *The Structure of Scientific Revolutions*. Chicago: University of Chicago Press, 1996.

Darwin, Charles. *The Voyage of the Beagle: Charles Darwin's Journal of Researches*. London: Penguin Classics, 1989.

Dillard, Annie. *Teaching a Stone to Talk: Expeditions and Encounters*. New York: Harper Perennial, 1988.

Thoreau, Henry David. *Walden and Civil Disobedience*. New York: Penguin Classics, 1986.

Darwin, Charles. *The Autobiography of Charles Darwin: 1809–1882*. Edited by Nora Barlow. New York: W.W. Norton, 1993.

Emerson, Ralph Waldo. *Nature and Selected Essays*. London: Penguin Classics, 2003.

Workman, Fanny Bullock and Workman, William Hunter. *Two Summers in the Ice-Wilds of Eastern Karakoram*. New York: E.P. Dutton, 1916.

"indescribably grand . . ." Longstaff, Thomas in "Glacier Exploration in the Eastern Karakoram," *The Geographical Journal* 35 (1910): 639.

Ali, Saleem. *Peace Parks: Conservation and Conflict Resolution.* Boston: MIT Press, 2007.

"We tell ourselves stories in order to live . . ." in Joan Didion, *The White Album.* New York: Farrar, Straus and Giroux, 2009.

PART TWO

4. *UNDERCURRENTS*

Ascherson, Neal. *Black Sea.* New York: Vintage, 2007.

Wood, Frances. *Did Marco Polo Go to China?* Boulder: Westview Press, 1996.

Calvino, Italo. *Invisible Cities.* New York: Houghton Mifflin Harcourt, 2013.

"heartbroken, wandering, wordless . . ." in Barks, Coleman. *The Essential Rumi.* New York: HarperCollins, 2010.

5. *THE COLD WORLD AWAKENS*

"'Trust me, there is order here . . .'" in Ondaatje, Michael. *In the Skin of a Lion.* Toronto: McClelland & Stewart, 2011.

"What is the colour of wisdom . . ." in Connell, Evan S. *Notes from a Bottle Found on the Beach at Carmel.* Berkeley: Counterpoint, 2013.

"polar exploration is . . ." in Cherry-Garrad, Apsley. *The Worst Journey in the World.* New York: Carroll & Graf, 1997.

Bildstein, Keith L. *Migrating Raptors of the World: Their Ecology and Conservation.* Ithaca: Cornell University Press, 2006.

Lilienthal, Otto. *Practical Experiments in Soaring.* Washington: Smithsonian Institution Annual Report, 1894.

"comical appearance of flying . . ." in Herlihy, David. *Bicycle: The History.* New Haven: Yale University Press, 2004.

"It was not uncommon . . ." in Crouch, Tom. *The Bishop's Boys: A Life of Wilbur and Orville Wright.* New York: W.W. Norton, 2003.

"Names are only the guests . . ." from Hsu Yu quoted in Domanski, Don. *All Our Wonder Unavenged.* London: Brick Books, 2007.

6. *ANGLE OF INCIDENCE*

Herlihy, David V. *Bicycle: The History*. New Haven: Yale University Press, 2004.

de Waal, Thomas. *The Caucasus: An Introduction*. Oxford: Oxford University Press, 2010.

Domanski, Don. *All Our Wonder Unavenged*. London: Brick Books, 2007.

"In the middle of the forest . . ." in Tranströmer, Tomas. *The Half-Finished Heaven: The Best Poems of Tomas Tranströmer*. Translated by Robert Bly. Minneapolis: Graywolf Press, 2001.

"keenest among the old at reading birdflight . . ." in Homer, *The Odyssey*. Translated by Robert Fitzgerald. New York: Farrar, Straus and Giroux, 1998.

7. *BORDERLANDIA*

"The creation of very particular human cultures . . ." in Cronon, William. *Uncommon Ground: Rethinking the Human Place in Nature*. New York: W.W. Norton, 1995.

"When we try to pick out anything by itself . . ." in John Muir, *Nature Writings*. New York: The Library of America, 1997.

"Each dying in its own way . . ." in Babel, Isaac. *The Collected Stories of Isaac Babel*. New York: W.W. Norton, 2002.

"Something there is that doesn't love a wall . . ." from "Mending Wall" in Frost, Robert. *The Poetry of Robert Frost: The Collected Poems, Complete and Unabridged*. New York: Henry Holt and Company, 2002.

"exactitude is not truth," in Matisse, Henri. *Matisse on Art*. Edited by Jack Flam. Berkeley: University of California Press, 1995.

PART THREE

8. *WILDERNESS/WASTELAND*

Berger, John. *Selected Essays*. New York: Knopf Doubleday, 2008.

"You cannot fill the Aral with tears . . ." Muhammad Salih as quoted in Weinthal, Erika. *State Making and Environmental Cooperation:*

Linking Domestic and International Politics in Central Asia. Boston: MIT Press, 2002.

Nelson, Craig. *Rocket Men: The Epic Story of the First Men on the Moon.* New York: Penguin, 2009.

Sagan, Carl. *Murmurs of Earth: The Voyager Interstellar Record.* New York: Random House, 1983.

Smith, Andrew. *Moondust: In Search of the Men Who Fell to Earth.* London: Bloomsbury, 2006.

Sagan, Carl. *Contact.* New York: Simon and Schuster, 1997.

Wallace, Alfred Russell. *My Life: A Record of Events and Opinions.* London: Chapman & Hall, 1908.

"Surely, for this great and holy purpose . . ." Alfred Russell Wallace's letter to *The Daily News*, February 6, 1909: 4, http://people.wku. edu/charles.smith/wallace/S670.htm.

"The day was lovely . . ." in Cheever, John. *Collected Stories and Other Writings.* New York: Library of America, 2009.

Lowell, Percival. *Mars and Its Canals.* New York: Macmillan, 1906.

Wallace, Alfred Russell. *Is Mars Habitable?* New York: Macmillan, 1907.

9. THE SOURCE OF A RIVER

Hopkirk, Peter. *The Great Game: The Struggle for Empire in Central Asia.* London: Hodder & Stoughton, 2006.

10. A MOTE OF DUST SUSPENDED IN A SUNBEAM

"Thus, a *wēijī* is indeed a genuine crisis . . ." Victor Mair, *Pīnyīn.info*, www.pinyin.info/chinese/crisis.html.

"What country's government would not protect . . ." Xu Jianrong, quoted in Michael Wines, "To Protect an Ancient City, China Moves to Raze It," *The New York Times*, May 28, 2009, www. nytimes.com/2009/05/28/world/asia/28kashgar.html.

"exposing the heinous reactionary . . ." and "All villages become fortresses, and everyone is a watchman," "China: No End to Tibet Surveillance Program," Human Rights Watch, www.hrw.org/ news/2016/01/18/china-no-end-tibet-surveillance-program.

"I am sometimes afraid . . ." in Darwin, Charles. *The Correspondence of Charles Darwin: 1821–1836,* vol. 1. Cambridge: Cambridge University Press, 1985.

"Look again at that dot . . ." in Sagan, Carl. *Pale Blue Dot: A Vision of the Human Future in Space.* New York: Random House, 1994.

Hinton, David. *Hunger Mountain: A Field Guide to Mind and Landscape.* Boston: Shambhala Publications, 2012.

Shakya, Tsering. *Dragon in the Land of Snows: The History of Modern Tibet since 1947.* New York: Columbia University Press, 1999.

Iyer, Pico. *The Open Road: The Global Journey of the Fourteenth Dalai Lama.* New York: Knopf Doubleday, 2008.

"The very same thought . . ." from Orgyen Tobgyal Rinpoche's teaching, August 17, 1999, www.rigpawiki.org/index.php?title=Lungta.

"the inhabitants of the city . . ." from van Schaik, Sam. *Tibet: A History.* New Haven: Yale University Press, 2011.

"The Tibetans saw giant 'birds' approach X . . ." Jianglin Li quoted in Siling, Luo, "A Writer's Quest to Unearth the Roots of Tibet's Unrest," *The New York Times,* August 14, 2016, www.nytimes.com/2016/08/15/world/asia/china-tibet-lhasa-jianglin-li.html.

11. *ROAD'S END*

"A mountain always practises in every place . . ." from Dōgen Zenji quoted in Snyder, Gary. *Practice of the Wild.* Berkeley: Counterpoint, 2004.

"absolutely unmixed attention . . ." in Simone Weil, *Gravity and Grace.* New York: Putnam, 1952.

"Maybe I will be able to go beyond my dream . . ." in Moitessier, Bernard. *The Long Way.* New York: Sheridan House, 1995.

Kate Harris is a writer with a knack for getting lost. Winner of the 2012 Ellen Meloy Desert Writers Award, her work has been featured in *The Walrus*, *Canadian Geographic Travel*, and *The Georgia Review*, and cited in *Best American Essays* and *Best American Travel Writing*. Named one of Canada's top modern-day explorers, her journeys edging the limits of nations, endurance, and sanity have taken her to all seven continents. She lives off-grid in Atlin, BC, as often as possible.

www.kateharris.ca